Winning Ways for Your Mathematical Plays, Volume 4

Winning Ways
for Your Mathematical Plays

Volume 4, Second Edition

Elwyn R. Berlekamp, John H. Conway, Richard K. Guy

A K Peters
Wellesley, Massachusetts

Editorial, Sales, and Customer Service Office

A K Peters, Ltd.
888 Worcester Street
Suite 230
Wellesley, MA 02482

Library of Congress Cataloging-in-Publication Data

Berlekamp, Elwyn R.
  Winning Ways for your mathematical plays / Elwyn Berlekamp, John H. Conway,
  Richard Guy.--2nd ed.
    p. cm.
  Includes bibliographical references and index.
  ISBN 1-56881-130-6 (v. 1) – ISBN 1-56881-142-X (v. 2) – ISBN 1-56881-143-8 (v. 3) – ISBN 1-56881-144-6 (v. 4) (alk.paper)
  1. Mathematical recreations. I. Conway, John Horton. II. Guy Richard K. III. Title.

QA95 .B446 2000
739.7'4--dc21   00-048541

Printed in Canada

08 07 06 05 04   10 9 8 7 6 5 4 3 2 1

To Martin Gardner

who has brought more mathematics to more millions than anyone else

Elwyn Berlekamp was born in Dover, Ohio, on September 6, 1940. He has been Professor of Mathematics and of Electrical Engineering/ Computer Science at UC Berkeley since 1971. He has also been active in several technology business ventures. In addition to writing many journal articles and several books, Berlekamp also has 12 patented inventions, mostly dealing with algorithms for synchronization and error correction.

He is a member of the National Academy of Sciences, the National Academy of Engineering, and the American Academy of Arts and Sciences. From 1994 to 1998, he was chairman of the board of trustees of the Mathematical Sciences Research Institute (MSRI).

John H. Conway was born in Liverpool, England, on December 26, 1937. He is one of the preeminent theorists in the study of finite groups and the mathematical study of knots, and has written over 10 books and more than 140 journal articles.

Before joining Princeton University in 1986 as the John von Neumann Distinguished Professor of Mathematics, Conway served as professor of mathematics at Cambridge University, and remains an honorary fellow of Caius College. The recipient of many prizes in research and exposition, Conway is also widely known as the inventor of the Game of Life, a computer simulation of simple cellular "life," governed by remarkably simple rules.

Richard Guy was born in Nuneaton, England, on September 30, 1916. He has taught mathematics at many levels and in many places—England, Singapore, India, and Canada. Since 1965 he has been Professor of Mathematics at the University of Calgary, and is now Faculty Professor and Emeritus Professor. The university awarded him an Honorary Degree in 1991. He was Noyce Professor at Grinnell College in 2000.

He continues to climb mountains with his wife, Louise, and they have been patrons of the Association of Canadian Mountain Guides' Ball and recipients of the A. O. Wheeler award for Service to the Alpine Club of Canada.

Contents

Preface to the Second Edition

This is the volume that has seen the fewest changes in the new edition, if we ignore typographical corrections and minor improvements of wording. Most of them take the form of additional references to the vast amount of work that has been done in our subject since the previous edition appeared.

Some more substantial corrections and additions have been made to the discussion of Rubik's Cube in Chapter 24, based on the ideas of several authors. We have also added a small section to that chapter to report on Marc Paulhus's solution to the question raised by one of us many years ago: "Can a game of Strip-Jack-Naked continue for ever?"

The Game of Life still has many adherents more than 30 years after its invention, and some of their discoveries and comments have been incorporated in Chapter 25.

We thank everyone who has written to us with suggestions for improving the book, even when we have not taken their advice. We also thank everyone at A K Peters who has worked on the book, in particular Jon Peters, who has fixed some of the illustrations. Additional thanks go to A and K themselves for having undertaken its republication without knowing what they were in for!

Elwyn Berlekamp, University of California, Berkeley
John Conway, Princeton University
Richard Guy, The University of Calgary, Canada

February 7, 2004

Preface to the Original Edition

Does a book need a Preface? What more, after fifteen years of toil, do three talented authors have to add.

We can reassure the bookstore browser, "Yes, this is just the book you want!"

We can direct you, if you want to know quickly what's in the book, to page xvi. This in turn directs you to volumes 1,2,3 and 4.

We can supply the reviewer, faced with the task of ploughing through nearly a thousand information-packed pages, with some pithy criticisms by indicating the horns of the polylemma the book finds itself on. It is not an encyclopedia. It is encyclopedic, but there are still too many games missing for it to claim to be complete. It is not a book on recreational mathematics because there's too much serious mathematics in it. On the other hand, for us, as for our predecessors Rouse Ball, Dudeney, Martin Gardner, Kraitchik, Sam Loyd, Lucas, Tom O'Beirne and Fred. Schuh, mathematics itself is a recreation. It is not an undergraduate text, since the exercises are not set out in an orderly fashion, with the easy ones at the beginning. They are there though, and with the hundred and sixty-three mistakes we've left in, provide plenty of opportunity for reader participation. So don't just stand back and admire it, work of art though it is. It is not a graduate text, since it's too expensive and contains far more than any graduate student can be expected to learn. But it does carry you to the frontiers of research in combinatorial game theory and the many unsolved problems will stimulate further discoveries.

We thank Patrick Browne for our title. This exercised us for quite a time. One morning, while walking to the university, John and Richard came up with "Whose game?" but realized they couldn't spell it (there are three tooze in English) so it became a one-line joke on line one of the text. There isn't room to explain all the jokes, not even the fifty-nine private ones (each of our birthdays appears more than once in the book).

Omar started as a joke, but soon materialized as Kimberly King. Louise Guy also helped with proof-reading, but her greater contribution was the hospitality which enabled the three of us to work together on several occasions. Louise also did technical typing after many drafts had been made by Karen McDermid and Betty Teare.

Our thanks for many contributions to content may be measured by the number of names in the index. To do real justice would take too much space. Here's an abridged list of helpers: Richard Austin , Clive Bach, John Beasley, Aviezri Fraenkel, David Fremlin, Solomon Golomb, Steve Grantham, Mike Guy, Dean Hickerson, Hendrik Lenstra, Richard Nowakowski, Anne Scott, David Seal, John Selfridge, Cedric Smith and Steve Tschantz.

No small part of the reason for the assured success of the book is owed to the well-informed and sympathetic guidance of Len Cegielka and the willingness of the staff of Academic Press and of Page Bros. to adapt to the idiosyncrasies of the authors, who grasped every opportunity to modify grammar, strain semantics, pervert punctuation, alter orthography, tamper with traditional typography and commit outrageous puns and inside jokes.

Thanks also the the Isaak Walton Killam Foundation for Richard's Resident Fellowship at The University of Calgary during the compilation of a critical draft, and to the National (Science & Engineering) Research Council of Canada for a grant which enabled Elwyn and John to visit him more frequently than our widely scattered habitats would normally allow.

And thank you, Simon!

University of California, Berkeley, CA 94720 Elwyn Berlekamp
University of Cambridge, England, CB2 1SB John Conway
University of Calgary, Canada, T2N 1N4 Richard Guy

You are
now here

If you want to know roughly what's elsewhere,
turn to the little notes about our four main themes:

Adding Games ... ... ♠ page 1
Bending the Rules ... ... ♡ page 277
Case Studies ... ... ♣ page 461
Doing It Yourself ... ... ◇ page 803

There are a number of other connexions between various chapters of the book:

However, you should be able to pick any chapter and read almost all of it
without reference to anything earlier, except perhaps the basic ideas at the start of the book.

Solitaire Diamonds!

Twinkle, twinkle little star,
How I wonder what you are!
Up above the world so high
Like a diamond in the sky!
     Jane Taylor, The Star.

We are all in the dumps, For diamonds are trumps;
The kittens are gone to St. Paul's.
the babies are bit, the Moon's in a fit,
And the houses are built without walls.
     Nursery Rhyme

If you've followed everyting in *Winning Ways* so far, you're probably finding it hard to get people to play with you, so you will need something to do on your own. Here are our favorite solitaire diamonds:

The classical game of Peg Solitaire, treated by old and new methods in Chapter 23.
A host of puzzles, pastimes and other party tricks in Chapter 24.
And finally, every automaton will enjoy playing the notorious game of Life (Chapter 25).

-23-

Purging Pegs Properly

We can merely mention bean-bags, peg-boards, size and form boards, as some of the apparatus found useful for the purpose of amusing and instructing the weak-minded.
Allbutt's Systematic Medicine, 1899, VIII, 246.

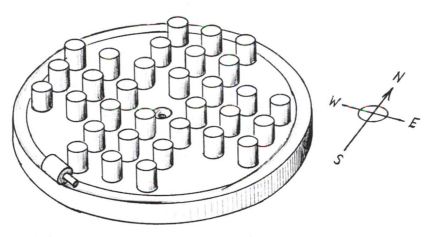

Figure 1. The English Solitaire Board.

Figure 1 shows the English Board on which the game of Peg Solitaire is usually played. It's easier to refill the board if you use marbles, but pegs are steadier when it comes to analysis.

The game is played (by one person of course) as shown in Fig. 2. If in some row or column two adjacent pegs are next to an empty space as in Fig. 2(a), then we may jump the peg p over r into the space s (Fig. 2(b)). The peg r that has been jumped over is then removed (Fig. 2(c)). Jumps are like captures in Draughts or Checkers, but they *never* take place diagonally, but only in the East, South, West or North directions.

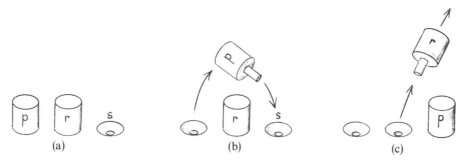

Figure 2. Making a Solitaire Jump.

Central Solitaire

The standard problem is to start as in Fig. 1, with a peg in every hole except the centre, and then aim, by making a series of these jumping moves, to reduce the situation to a single peg in the central hole (Fig. 3).

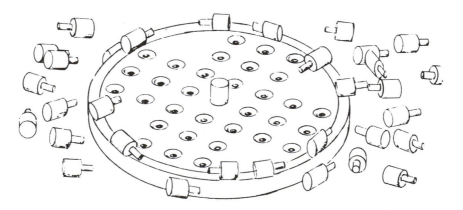

Figure 3. Success!

Like many card solitaire ("Patience") games, Solitaire is probably called a *game* rather than a puzzle because one often feels one is playing against an invisible opponent. Many people not normally interested in puzzles will recall some period of their lives when they have struggled with this opponent for days at a time; yet it seems that most of those who can readily solve simple Solitaire problems have been taught the trick by someone else as a child. It is rare indeed to find someone who has acquired the knack single-handed, and surely Peg Solitaire (nowadays selling in many parts of the world under the trade name of Hi-Q) must be the hardest game of its kind to have gamed substantial popularity. It is an ideal game to while away hours of enforced idleness during illness or long journeys, and perhaps we should believe those old books which tell us that the game was invented by a French nobleman who first played it on the stone tiles of his prison cell.

If you haven't played this game before, put down this book, go out right now, buy a board, and try to solve the Central Solitaire Game. Those of you who are left will have plenty of time to read the chapter before the novices come back in a week or so—why not learn a particularly elegant solution to impress them all?

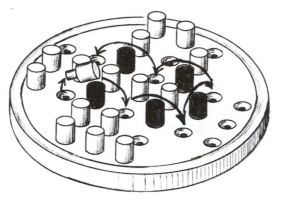

Figure 4. A Move of Five Jumps.

Dudeney, Bergholt and Beasley

Since you must already know how to solve the problem, you'll want to do it quickly, so let's agree to count any number of consecutive jumps with a single peg as just one **move**. Figure 4 shows such a move—the five shaded jumped-over pegs are to be taken off as part of the move.

			a	b	c		
		y	d	e	f	z	
	g	h	i	j	k	l	m
	n	o	p	x	P	O	N
	M	L	K	J	I	H	G
		Z	F	E	D	Y	
			C	B	A		

Figure 5. Labelling the Places.

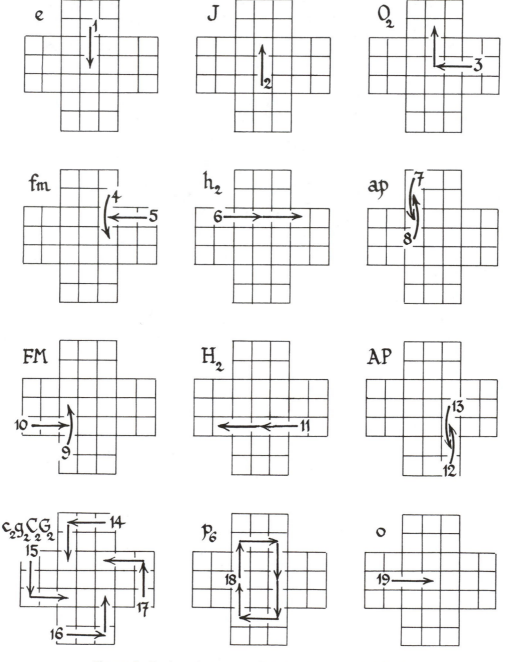

Figure 6. Dudeney's 19-move Solution for Central Solitaire.

In order to describe a solution concisely, we label the places as in Fig. 5, and write S_t for a jump from S to t and shorten this to S when we don't need to indicate the direction. The 5-jump move of Fig. 4 is L_{JHljh} which we will abbreviate to L_5 when it is unambiguous (we can't do that here since L_5 could also mean L_{hjJHl} and various other things). In this notation, Dudeney's elegant 19-move solution of his Central Solitaire problem is

$$eJO_2 \; fmh_2ap \; FMH_2AP \; c_2g_2C_2G_2 \; p_6o,$$

and this is set out in Fig. 6.

Dudeney thought that the number 19 could not be improved, but, in *The Queen* four years later, Ernest Bergholt gave an 18-move solution, unfortunately not quite as symmetrical as Dudeney's:

$$elcPDGJm_2igL_5CpA_2M_2a_3d_5o.$$

Here the notation L_5 is ambiguous, but the intended 5-jump move is the one depicted in Fig. 4. The move d_5 is also ambiguous, but either interpretation leads to the same result.

The whole truth emerged only 52 years later, in 1964, when John Beasley used the methods described in this chapter to prove that a solution in fewer than 18 moves is impossible. With Beasley's kind permission we publish his proof for the first time in the Extras to this chapter. It is very condensed, so the reader who wishes to follow it should first study the chapter diligently!

Packages and Purges

It's nice to be able to know the effect of a whole collection of moves before you make them, so let us sell you some of our instant **packages**. When a package is used to clear all the pegs from a region, we call it a **purge**.

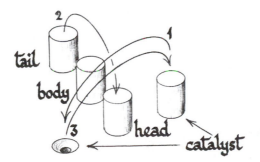

Figure 7. Purging Three Pegs.

Figure 7 shows the handy little **3-purge**, our most popular package. When three pegs—the *tail*, the *body* and the *head*—are adjacent in line, this will remove them all, provided the head has an additional peg on one side of it, and an empty space on the other, as in the figure. Move 1 of the package jumps the additional peg *over* the head; move 2 jumps tail over body *into* head; and move 3 jumps back *over* the head to its original position. Since the peg and the space on either side of the head are essential to the package, but are restored to their original state, we call them **the catalyst**.

In Figs. 8(a) to 8(h), ● indicates a peg to be purged, ○ a space to be filled, and ✕✕ indicate catalyst places of which one must be full and the other empty. In most of the purges there are two catalyst moves in opposite directions over the same position (which may initially be either a peg or an empty space) and the remaining moves form one or two packages which deliver pegs to that place. For the 3-purge (8(a)), one peg was already in place and the second is delivered by a single jump which we might call the "2-package" (8(b)). The 6-purge is usually accomplished (8(c)) using a 2-package to deliver the first peg and a 4-package (8(d)) for the second.

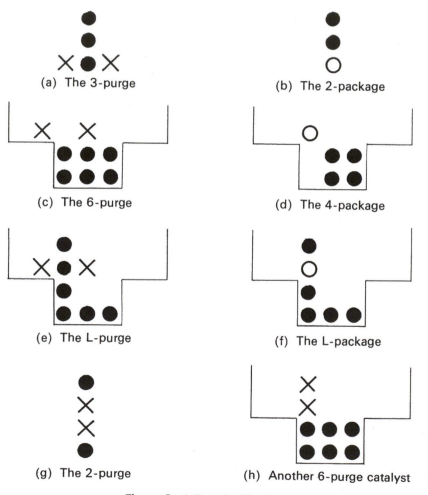

Figure 8. A Parcel of Packages.

The L-purge (8(e)) and L-package (8(f)) are very useful indeed. The first peg for the L-purge is already in place and an L-package supplies the second. The first two moves of the L-package form a 2-purge (8(g)) which can also be used in other situations. The catalyst for the 2-purge is restored in a rather unorthodox way, as is the alternative catalyst for the 6-purge shown in Fig. 8(h).

Packages Provide Perfect Panacea

Plenty of problems are performed with panache by people who purchase our packages.

In Fig. 9 we can see at a glance a solution for Central Solitaire, consisting of two 3-purges (1 and 2) followed by three 6-purges (3, 4 and 5) and an L-purge , leaving only the final jump to be made. You should check that every purge has the catalyst it needs.

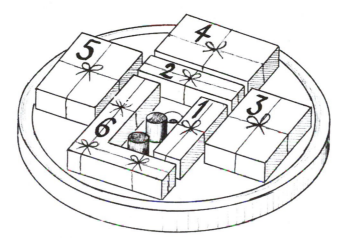

Figure 9. Central Solitaire Painlessly Packaged.

Instead of Central Solitaire we can consider other one-peg reversal problems: start with only one empty space and finish with only one peg in the same place. Figures 9 and 10 show that most such problems can be solved by purely purgatory methods, but in Fig. 10(e) we start with a 4-package indicated by the arrow (1), and the notorious problem (b) needs more complicated methods.

To clarify our notation we explain our solution for (b) in detail. For the first jump we have no choice but to jump from the place marked 1 in the figure. Our second jump, from the place marked 2, clears a space which enables us to make the L-*package*, indicated by the bent arrow (3). We now have a catalyst for the L-purge (4) which is followed by a single jump from the place marked 5. We are now on the home run with purges 6, 7 and 8 followed by a single jump from place 9. If the reader plays this through she will find that we have set up a spectacular 5-jump move from the place marked 10_5.

The reader might like to try her hand at some *two*-peg reversal problems—start with just two spaces on the board and end with just two pegs in those places.

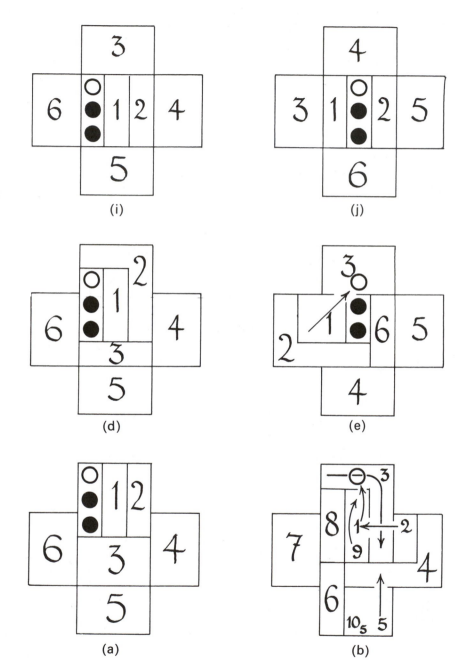

Figure 10. The Other Six One-Peg Reversals.

The Rule of Two and the Rule of Three

Here is another type of problem (Fig. 11). We start with just one empty space and declare that some particular peg is to be the **finalist** (last on the board). In the example the initial hole is at position d and we want the finalist to be the peg that starts at b. Where must it end?

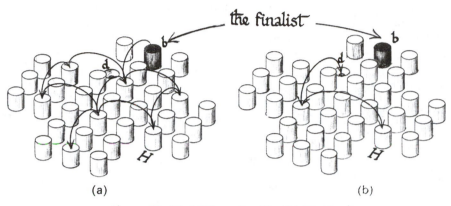

Figure 11. Find Where the Finalist Finishes!

There is an obvious **Rule of Two**—the peg can only jump an even number of places in either direction, as indicated by the arrows in Fig. 11(a). But there is a much more interesting **Rule of Three**. One of the consequences of this is that if we start with a single space on the English board and end with a single peg, then we can move in steps of three from the initial space to that of the finalist, as in Fig. 11(b).

The Rule of Two and the Rule of Three, taken together, can lead to surprises. See how they point to the unique finishing place H in Fig. 11(a) and (b). Now that we know that H is the only place permitted by both the Rule of Two and the Rule of Three, the problem is a lot easier than it might have been. Figure 12 shows a neatly packaged solution; how did we find it?

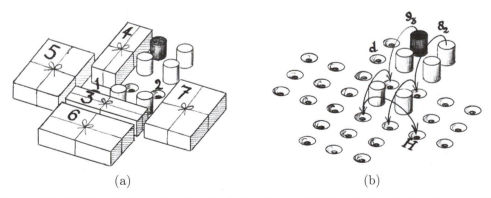

Figure 12. (a) The Position after the First Two Moves. (b) The Position before the Last Two Moves.

What we did was plan the 3-jump move 9_3 which puts the finalist in his place, and our second jump was to clear a space for this. But after we made this second jump most of the pegs parcelled themselves up naturally. The one apparent exception was the peg starting just right of the finalist, and the best way of clearing this seemed to be to use it as in move 8_2 to provide the final jump.

For other problems, gentle reader, we recommend a similar procedure. Plan the last few moves of your solution and let the first few be used to smooth the way for these and leave the remaining pegs in tidy packages. Remember that the catalyst for the very last purge must be among the pegs in your planned finale.

Here's a nice finalist problem for you. Let the initial hole be in position B and the finalist be the peg which starts at J. Can you end with only this peg?

Some Pegs Are More Equal Than Others

How do we explain the Rule of Three? The best way is to introduce "multiplication" for Solitaire positions. In Fig. 13(a) the two adjacent pegs s and t can obviously be replaced by a single peg at r, so we write

$$st = r,$$

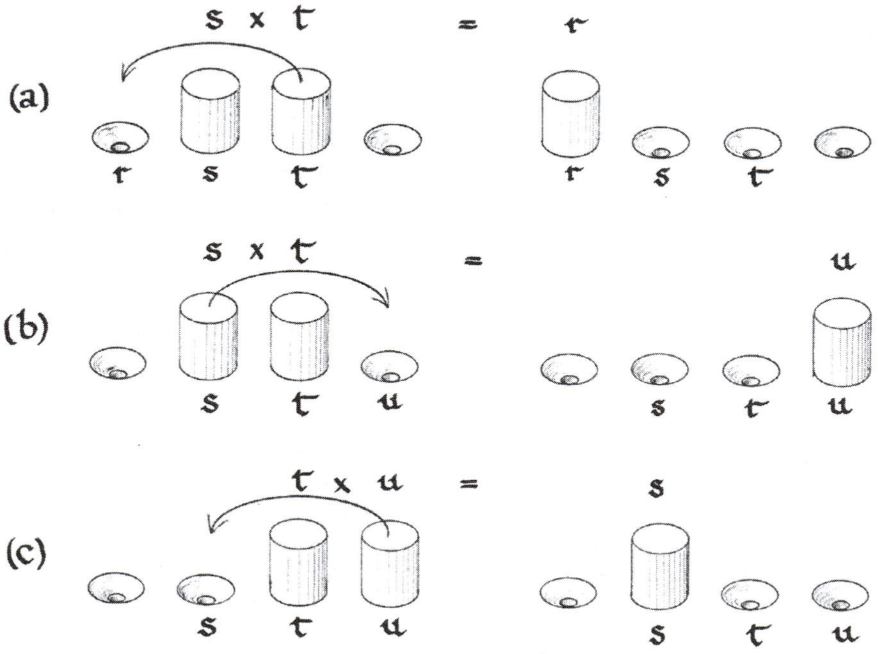

Figure 13. Multiplying Pegs.

but Fig. 13(b) shows that we can also write

$$st = u.$$

Now Euclid tells us that things that are equal to the same thing are equal, so we must agree that $r = u$.

> Places three apart
> in any line are
> considered equal.

Let's see what other rules of algebra tell us. Combining Figs. 13(b) and 13(c), we have

$$st = u, \ \ tu = s,$$
$$st^2u = us,$$

or, cancelling,

$$t^2 = 1,$$

which seems to tell us that

> two pegs in the
> same place cancel.

Remember how catalysts do precisely this—they remove two pegs which are delivered to the same place by the other moves of a purge. In fact it follows from our algebra that

> any set of pegs
> that can be
> purged cancel.

For example, in Fig. 13(c), $tu = s$, so

$$stu = ss = 1.$$

> Three adjacent pegs   (3-purge)
> in line cancel,

and since $r = u$,

$$ru = uu = 1.$$

$$\boxed{\begin{array}{c} \text{Two pegs at distance} \\ \text{three cancel.} \end{array}} \qquad \text{(2-purge)}$$

But in the algebra there are less obvious equalities: for since $s^2 = 1 = rst$, we find

$$s = rt.$$

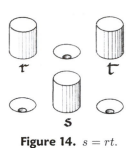

Figure 14. $s = rt$.

Reiss's 16 Solitaire Position Classes

We've now said enough to see how our algebra cuts the Solitaire board down to size, for since places three apart are algebraically equal, every place is equal to one of the nine in the middle of the board (Fig. 15); for example $a = p$. Now we can use our most recent rule to express each of these nine in terms of the four corner ones, i, k, I, K:

$$\begin{array}{ll} j = ik & P = Ik \\ p = iK & J = IK \\ x = jJ = ikIK. & \end{array}$$

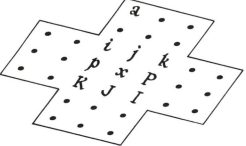

Figure 15. Stripping Down to Essentials.

Since equal pegs cancel,

> every Solitaire position
> is algebraically equal to
> one of the 16 combinations
> of the places i, k, I, K.

THE SIXTEEN REISS CLASSES

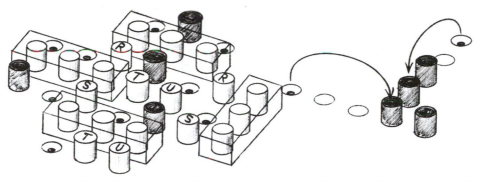

Figure 16(a). Found near Split?   **Figure 16 (b).** Reduced to Size.

The position of Fig. 16(a) was found unattended in a Jugoslav railway train. Those filmy packages and letters weren't there—but just came into our mind's eye when we pondered the possibility of reducing the position to a single peg. Where must this single peg be?

Our rules allow us to cancel those four packages of three and then the four pairs RR, SS, TT, UU, so that the position is algebraically equal to the four shaded pegs. We then can move two of these three spaces and cancel another 3-package as in Fig. 16(b) to see that the position equals a single peg at I. So the Rule of Three says that the finalist must be at I, L or f. For which of these places can you find solutions?

How do we know that Reiss's sixteen classes are really different? Might not our algebraic rules imply perhaps that $i = kK$? No! For consider the numbers ± 1 shown in the places of Fig. 17(i). Whenever three of these numbers

$$r, s, t$$

are adjacent in line, we really do have

$$rs = t,$$

and from this we can see that all our algebraic rules hold for these numbers. But in this system we have

$$i = -1, \quad k = K = +1,$$

so we can't prove $i = kK$! In fact Figs. 17(i, k, I, K) show that all 16 combinations of the pegs i, k, I, K are algebraically distinct: for example the value on Fig. 17(i) is -1 just if i is involved in the combination. Making a Solitaire move or applying any of our algebraic rules will never change the value in any of the four Figures.

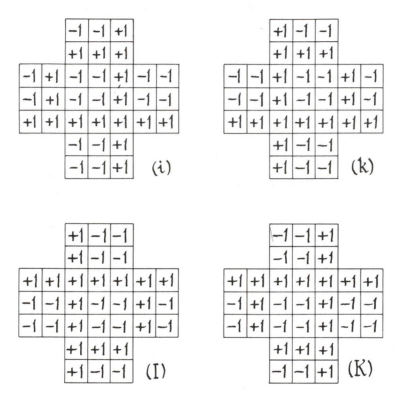

Figure 17. "Answers" to the Algebra.

In algebraic language, the first thing we told you about the Rule of Three may be restated— a position with just one empty space is algebraically equal to the **complementary** position in which *only* that place is full. More generally,

> any position on the English board is algebraically equal to the complementary position which has empty spaces replacing pegs and pegs replacing empty spaces.

For our rules allow us to complement any line of three adjacent places, and the whole board can be parcelled into such threes.

This property fails for the Continental Board.

The Continental Board

The Continental Board which has the four extra holes at y, z, Y, Z in Fig. 5. So no reversal problems are possible on this board. Which of the problems which start with a single hole and end with a single peg are solvable on this board? See the Extras.

Playing Backwards and Forwards

> "The game called Solitaire pleases me much. I take it in reverse order. That is to say that instead of making a configuration according to the rules of the game, which is to jump to an empty place and remove the piece over which one has jumped, I thought it was better to reconstruct what had been demolished, by filling an empty hole over which one has leaped."

Leibniz.

The famous philosopher plainly thought that playing Solitaire backwards was different from playing it forwards, but really it's exactly the same game! For let's see what happens when he makes one of his backward moves from Figs. 18(a) to 18(c). Leibniz regards this as jumping *piece t* into *hole r* and *filling* the empty *hole s* over which he has leaped, but Fig. 18(b) shows that we can regard him as jumping the *hole* at r over the *hole* at s into the *piece* at t and removing the *hole* over which he has jumped. (Of course to *remove* a hole he *inserts* a piece!)

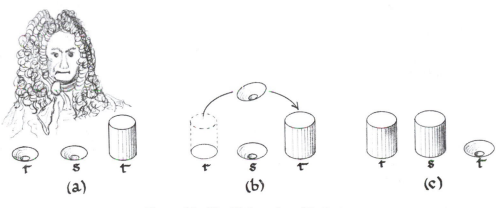

Figure 18. The Philosophy of Leibniz.

> Backwards Solitaire is just forwards Solitaire with the notions "empty" and "full" interchanged.

TIME-REVERSAL = ANTI-MATTER?

This can be useful as well as interesting. A quite spectacular Solitaire finale happens in Beasley's remarkable 16-move solution of the i-reversal problem:

$$apc_2F_2gdM_2IAP_\downarrow f_\downarrow C_3Gm\ldots.$$

After 14 of the 16 moves the board still seems quite full (Fig. 19(a)) but can be cleared to a single peg in just 2 moves. (Can you find them?)

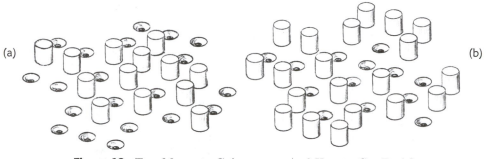

Figure 19. Two Moves to Go!   ...   And How to Get Back?

How would *you* find the moves leading up to this positron? The time-reversal trick should make it easy. Instead of reducing a position with only one space, at i, to Fig. 19(a), try to reduce the complementary position (Fig. 19(b)) to just one peg at i. If you've been doing your homework and practising diligently, you won't find this too hard. You too can astonish your friends with grand finales to other Solitaire problems set up by the time-reversal trick.

Pagoda Functons

Reiss's algebraic theory (known to many!) applies even when we allow you to make moves backward in time (like Leibniz) as well as the ordinary forward ones. Of course this lets you take back any of your bad moves, but you may also "undo" moves you haven't even made! If two positions are in different Reiss classes, then we can never get from one to the other by normal moves, by Leibniz's backward moves nor by any mixture of the two.

Unfortunately this means, of course, that the Reiss theory can never tell you when you've made a bad move, because the Reiss class never changes. You need something like the **Pagoda Functions** (known to few!) we are about to show you, that can *change* when you make a move, albeit in a restricted way. Mike Boardman was one of those who helped us to develop these.

Those friends of yours should now be back from the store with their Solitaire boards, so why not present them with a couple of innocent-looking problems? Since these are *reversal* problems, your friends won't be able to prove them impossible even if they've got as far as the last section.

The two problems are shown in Figs. 21(a) and (b) where circles show the only places which are initially empty and which must also be the only places which are finally full. Figures 21(c) and (d) show two pagoda functions which prove the problems impossible. In general, if pag is any such function and X any Solitaire position, we shall write

$$\text{pag } X$$

for the *sum* of the numbers that pag assigns to the pegs which are present in X. If X is partitioned into smaller positions Y and Z, then, in our algebraic notation we have

$$X = YZ,$$

and

$$\text{pag } X = \text{pag } Y + \text{pag } Z.$$

so that pagoda functions behave like logarithms.

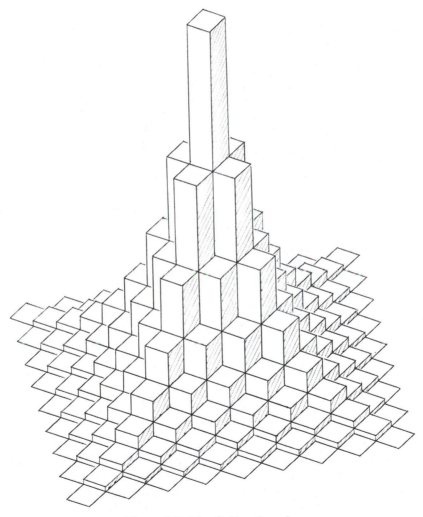

Figure 20. The Golden Pagoda.

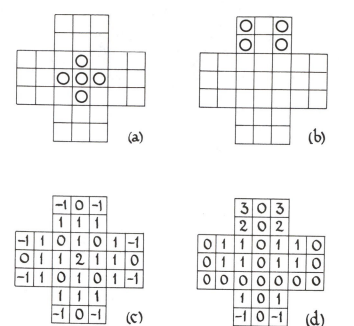

Figure 21. Two Impossible Reversal Problems.

The essential property which defines pagoda functions is that *no move may increase the value.* To check this condition you must make sure that

THE PAGODA FUNCTION CONDITION

$$\text{pag } r + \text{pag } s \geq \text{pag } t$$

holds for every conceivable Solitaire jump r over s into t.

CHECK THIS CONDITION NOW IN FIGURES 21(c) AND (d)!

When you've done that you will see the impossibility of our two problems, since the pag (Fig. 21(c)) of the initial position in 21(a), namely 4, can't be increased to 6, the pag of the final position; nor can 8 be increased to 10 (Figs. 21(b) and (c)).

In Fig. 22 we show the pagoda functions you're most likely to find useful; so you'd better check the Pagoda Function Condition for each of them! The values in the blank spaces are zero and you can make any of the indicated swaps. Figures 22(c, d, h, and v) are obvious pagoda functions since they just indicate all the places that a given peg can go to. The 12 places in 22(c) are called **corners** and the 5 in 22(d) are the **dodos**, because one of the easiest mistakes you can make is to let your dodos become extinct when you need one in your final position. Those extra minus ones often make 22(a) and (b) more useful than (h) and (v).

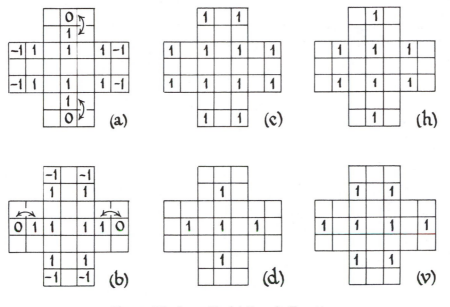

Figure 22. Some Useful Pagoda Functions.

The Solitaire Army

A number of Solitaire men stand initially on one side of a straight line beyond which is an infinite empty desert (Fig. 23). *How many men do we need to send a scout just 0, 1, 2, 3, 4, or 5 paces out into the desert?*

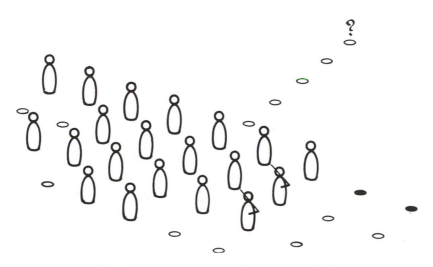

Figure 23. How About Sending a Scout Out?

Chapter 23. Purging Pegs Properly   ◇

It's not hard to see that the answers for 0, 1, 2 and 3 paces are 1, 2, 4 and 8 men, so you might guess that the next two answers are 16 and 32. But in fact no less than 20 men are needed to get 4 paces out. Can you find the two possible configurations of 20 men? (See the Extras.)

For 5 paces the answer is even more surprising—it is *impossible* to send a scout five paces into the desert , no matter how large an army we hire! The pagoda function which proves this is shown in Fig. 24. It was the shape of the graph of this function (Fig. 20) which first suggested the name "pagoda". The number σ is determined by the golden ratio:

$$\sigma \;=\; \frac{1}{2}(\sqrt{5}-1) = 0\cdot 618\ldots,$$
$$\sigma^2 + \sigma \;=\; 1.$$

$$1$$
$$\sigma$$
$$\sigma^2$$
$$\sigma^4 \quad \sigma^3 \quad \sigma^4$$
$$\sigma^6 \quad \sigma^5 \quad \sigma^4 \quad \sigma^5 \quad \sigma^6$$

$$\sigma^{10} \quad \sigma^9 \quad \sigma^8 \quad \sigma^7 \quad \sigma^6 \quad \sigma^5 \quad \sigma^6 \quad \sigma^7 \quad \sigma^8 \quad \sigma^9 \quad \sigma^{10}$$
$$\sigma^{11} \quad \sigma^{10} \quad \sigma^9 \quad \sigma^8 \quad \sigma^7 \quad \sigma^6 \quad \sigma^7 \quad \sigma^8 \quad \sigma^9 \quad \sigma^{10} \quad \sigma^{11}$$
$$\ldots \quad \sigma^{11} \quad \sigma^{10} \quad \sigma^9 \quad \sigma^8 \quad \sigma^7 \quad \sigma^8 \quad \sigma^9 \quad \sigma^{10} \quad \sigma^{11} \quad \ldots$$
$$\ldots \quad \sigma^{11} \quad \sigma^{10} \quad \sigma^9 \quad \sigma^8 \quad \sigma^9 \quad \sigma^{10} \quad \sigma^{11} \quad \ldots$$
$$\ldots \quad \sigma^{11} \quad \sigma^{10} \quad \sigma^9 \quad \sigma^{10} \quad \sigma^{11} \quad \ldots$$

$$\ldots\ldots\ldots\ldots\ldots\ldots\ldots\ldots$$
$$\ldots\ldots\ldots\ldots\ldots\ldots\ldots\ldots\ldots$$

$$\sigma^{n+5} \quad \sigma^{n+4} \quad \sigma^{n+3} \quad \sigma^{n+2} \quad \sigma^{n+1} \quad \Big|\, \sigma^n \quad \sigma^{n+1} \quad \sigma^{n+2} \quad \sigma^{n+3} \quad \sigma^{n+4} \quad \sigma^{n+5}$$

$$\ldots\ldots\ldots\ldots\ldots\ldots\ldots\ldots\ldots\ldots\ldots\ldots\ldots\ldots\ldots$$

Figure 24. Pagoda Function for the Solitaire Army.

By some easy mathematics we have

$$\sigma^n + \sigma^{n+1} + \sigma^{n+2} + \ldots = \frac{\sigma^n}{1-\sigma} = \sigma^{n-2},$$

so that the total score of the line whose middle element is σ^n is

$$\sigma^{n-2} + \sigma^{(n+1)-2} = \sigma^{n-3},$$

and the total score of this line and all lower lines is

$$\frac{\sigma^{n-3}}{1-\sigma} = \sigma^{n-5}.$$

In particular, the sum of *all* the men on or below the σ^5 line is *exactly* 1, so no finite number of these men will suffice to send a scout to the place whose score is 1. But infinitely many men are *almost* enough, because we once showed that if any man of our army is allowed to carry a comrade on his shoulders at the start, then no matter how far away the extra man is, the problem can now be solved.

Managing Your Resources

Your score on a pagoda function is in some sense a measure of your resources, which you should not consume too rapidly. But mere worldly goods are not enough: they must be capably managed to preserve a balance between your commitments in various directions.

The **Balance Sheet** of Fig. 25 has been cunningly devised to do just this. The subtlety of the English board is that you are often forced to consume assets in order to maintain the

Figure 25. The Balance Sheet.

balance, as measured by the greek letters α and β, of your position in the North-South and East-West directions. The latin letters a, b and c measure the **assets** on a number of pagoda functions simultaneously (a and b for Figs. 22(a) and 22(b) and abc^2 for Fig. 21(c)).

To estimate the overall capacity of a position, find the product of the resources of all its pegs in Fig. 25, using the relations

$$\alpha^2 = \beta^2 = 1.$$

A problem has two such products, the **raw product** (for its initial position), which must be taken to the **finished product** (for the final position) while consuming the **available resources**:

$$\frac{\text{raw product}}{\text{finished product}} = \text{available resources}.$$

In Fig. 26, all the jumps that change the product are shown to do so in **units** of sizes

$$
\begin{array}{ccccccc}
a & a\alpha & c\alpha & a^2c^{-1}\alpha & = A & a^2 \\
b & b\beta & c\beta & b^2c^{-1}\beta & = B & b^2
\end{array}
$$

so your available resources will only be **productive** if they can be made up of such units.

Central Solitaire, for example, has raw product a^4b^4 and finished product $c\alpha\beta$ so that its available resources are

$$\frac{a^4b^4}{c\alpha\beta} = a^4b^4c^{-1}\alpha\beta.$$

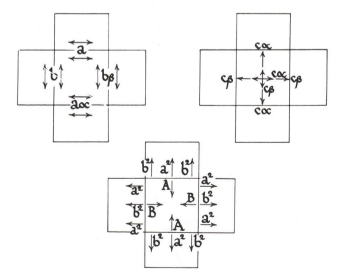

Figure 26. Using Resources in Various Units.

In Dudeney's solution only the opening and closing few moves actually use any of these:

move	e	J	O_2	$fmh_2apFMH_2APc_2g_2C_2G_2$		P_6	o
resources	A	$c\beta$	$B.c\alpha$	$\longleftarrow$ free moves $\longrightarrow$		$1.a.1.1.a\alpha.1$	B

Unproductivity

Many problems are impossible for the simple reason that

$$\boxed{\begin{array}{c} b^2\alpha \text{ and } a^2\beta \text{ are} \\ \text{unproductive!} \end{array}}$$

Why is this? In the case $b^2\alpha$, for example, we are hamstrung for lack of a's, so the α forces us to make a jump $c\alpha$, leaving only b^2c^{-1} for the remaining moves, in which c^{-1} demands a move $b^2c^{-1}\beta = B$, and we then have no assets with which to adjust the remaining β.

The Prodigal Son's Opening

$$\boxed{\begin{array}{c} \text{Jump into centre; jump over centre;} \\ \text{jump into centre; jump back over centre;} \end{array}}$$

is the only way Central Solitaire can go wrong in as few as four moves. What's so bad about these moves? The prodigality lies in the second and fourth moves which both use $c\alpha$ or both use $c\beta$ and therefore leave only

$$a^4b^4c^{-1}\alpha\beta/c^2 = a^4b^4c^{-3}\alpha\beta$$

for the remaining moves. But

$$\boxed{\begin{array}{c} a^4b^4c^{-3}\alpha\beta \\ \text{is unproductive!} \end{array}}$$

For the only way to cope with c^{-3} without overspending either a or b is to use the units

$$A, A, B \text{ or } A, B, B$$

which leave only the unproductive products

$$b^2\alpha \text{ or } a^2\beta.$$

Of course the same argument shows that *no* two moves in *any* solution of Central Solitaire can have product c^2.

Can you find the only way (**Fool's Solitaire**) of getting absolutely pegbound (unable to move) in six jumps? And can you **Succour the Sucker** by solving the position reached after five of these moves? And can you flag yourself down to *another* pegbound position in as few as ten jumps from the start? See the Extras.

Deficit Accounting and the G.N.P.

The **deficit** of a problem is the amount by which its initial position lacks the resources of the entire board, combined with the total resources of the final position and the costs of any moves you intend to make. Since the resources of the entire board are

$$a^4 b^4 c \alpha \beta \text{(the (English) Gross National Product)}$$

we have

$$\text{remaining resources} = \frac{a^4 b^4 c \alpha \beta}{\text{deficit}}.$$

The deficit is found very easily by multiplying the initial hole values by the final peg values. For Central Solitaire, the basic deficit is

$$c \alpha \beta . c \alpha \beta = c^2,$$

which the Prodigal Son's bad moves extravagantly enlarged to c^4. He clearly didn't know the **Deficit Rule**:

If deficit/c^4 *IS* productive,
your remaining resources AREN'T!

This is because (G.N.P.)/c^4 is our unproductive product $\alpha^4 \beta^4 c^{-3} \alpha \beta$.

Accounting for Two-Peg Reversal Problems

We know that all the one-peg reversal problems are possible, but there are just four different impossible two-peg reversal problems. The first of these is **Hamlet's Memorable Problem** (to be or not to be):

Get to only b, e present   (to be)
from only b, e absent   (not be).

Deficit account for Hamlet's Problem

To:

Initial holes @ b & e :	$\beta.\alpha\beta = a$
Final pegs @ b & e :	$\beta.\alpha\beta = a$
First & last jumps into e :	$c\alpha.c\alpha = c^2$
Jump into b :	a^2
Deficit :	$a^4 c^2$

Since

$$\frac{a^4 c^2}{c^4} = a^4 c^{-2} = A^2$$

is productive, Hamlet's Problem succumbs to the Deficit Rule. The other three impossible two-peg reversals are the **Dodo Problems**, for which the two places are two of the five dodo pegs (Fig. 22(d)). Deficit accounts for the typical problems eo, ex and eE are:

Dodo Problem	eo	ex	eE
Initial holes and final pegs $\Big\}$	$(\alpha\beta.b\alpha)^2$	$(a\beta.c\alpha\beta)^2$	$(a\beta.a\alpha\beta)^2$
Required moves	$c\alpha.c\beta$	$c\alpha$	$c\alpha.c\alpha$
Deficit	$a^2 b^2 c^2 \alpha\beta$	$a^2 c^3 \alpha$	$a^4 c^2$
Deficit/c^4	$A.B$	A	$A.A$

The reader who has been paying attention will have no difficulty in finding solutions to any other two-peg reversal problem.

John Conway, Mike Guy and Bob Hutchings have shown that the only impossible *three*-peg reversals are typified by

1. The **Bumble-bee Problems** (b, e and any third place other than g, m, M, G),

2. The **Deader Dodo Problems** (two dodos and any third place other than an outside corner $acgmMGCA$),

3. The **Three B'ars Problems** (any three of the unlucky 13 places in the three rows def, $nopxPON, FED$).

These can be shown to be impossible by deficit accounting . In fact in any reversal problem, an additional peg other than an outside corner merely aggravates the deficit.

Forgetting the Order Can Be Useful

If you allow yourself to have 2 or more, or -1 or less, pegs in a hole, you can make your moves in any order! It's a good idea to alter a hard problem in this way, and when you've solved the altered problem, go back and find a sensible order for the original one.

We'll do out loud for you the tricky 3-peg reversal:

start   with 0 pegs in b, N, n;   1 peg everywhere else;
end   with 1 peg in b, N, n;   0 pegs anywhere else.

In the altered problem it's easier to

start   with -1 peg in b, N, n;   1 peg everywhere else;
end   with 0 pegs anywhere.

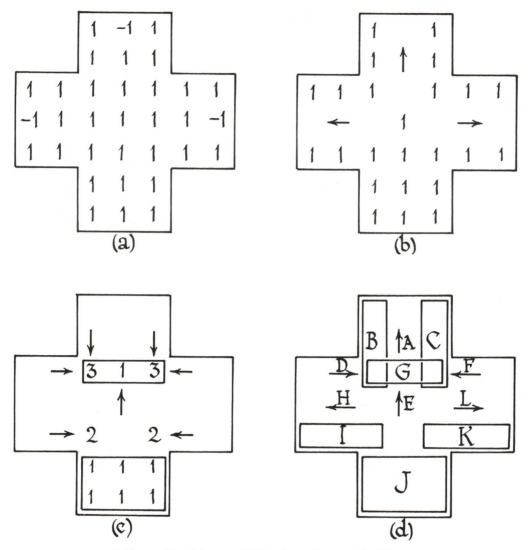

Figure 27. Solving a Tricky 3-peg Reversal Problem.

From the starting position, Fig. 27(a), we'll need, at some time, to make three jumps to fill those −1's, so we make these three jumps *now*, reaching Fig. 27(b). The only plausible way to deal with the six isolated corner pegs in this is to jump them inwards, and after the indicated upwards jump over the centre we reach Fig. 27(c). The remaining pegs in this can be cleared by a 3-purge, a 6-purge and four double jumps over the inner corners.

If you follow Fig. 27(d) in the order A to L you'll find yourself making all the above moves in a legal way. The double jumps have been incorporated into the 3-purges B, C, I and K.

Beasley's Exit Theorems

Sometimes you can work out exactly what moves to make in a problem, but find it hard to get them into the right order. The following remarks can help you get your moves in order, or prove that it can't be done.

> A region of at least three squares
> that starts full *or* ends empty
> needs at least one *exit move*.
> A region of at least three squares
> that starts full *and* ends empty
> needs at least two exit moves.

BEASLEY'S FIRST AND SECOND EXIT THEOREMS

An **exit move** for a region is a jump that empties some square in the region and fills some square outside the region. To justify Beasley's Second Exit Theorem, note that the first and last moves affecting a region must both be exits. We'll illustrate with a stolid survisivor problem.

A Stolid Survivor Problem

Suppose we want to do an a-reversal, with the added condition that peg K is the **stolid survivor**, i.e. that the first move of K is also the final move from K to a. Can the grand finale be a 6-chain?

The ideas of the first part of our discussion are often useful in long chain problems. Then we'll try to put the moves we've found into order, using Beasley's Exit Theorems.

How do we use the 16 side pegs $h^8 v^8$ of Figs. 22(h) and (v)? Each of the outer corner pegs must at some time be jumped into the central 3×3 square, and those at C and M must sidestep first to avoid the stolid survivor at K. So the jumps mentioned use up side pegs as follows:

c	m	G	A	C	M	g
v	h	h	v	hv	vh	h

leaving $h^3 v^4$ for the remaining jumps. Since the first move uses one side peg and the final chain six, we have accounted for all the side pegs and no *other* move can destroy one.

This forces us to make the first move c_a, since the alternative, i_a, would move an inside corner peg to the outside and make us use another side peg to bring it back later. Next k_c would use another side peg, so the second move is j_b and this peg must stay at b until the grand finale because a move refilling e would use yet another side peg. We now know that the final 6-chain uses $h^2 v^4$ and involves a horizontal jump over b, so it must be as in Fig. 28(a).

Since K doesn't move till the end, L can't be jumped over and can only be cleared by the upward jump L_h. We need to make two jumps over B, once to get corner peg C out,

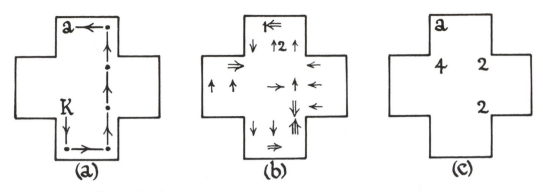

Figure 28. Can the Stolid Survivor Make a Grand 6-chain Finale?

and once in the finale, so we must deliver an extra peg there by a downward jump J_B. For similar reasons *two* extra pegs are needed at D, so we must make two downward jumps P_D. For the second of these, and for the finale, we need two more pegs delivered at P; these must come from N and p. We've now found 23 (Fig. 28(b)) of the 31 jumps. If we make these we arrive at Fig. 28(c). The two pegs on each of I and k must be cleared by pairs of vertical or horizontal to and fro jumps, and the four on i by two such pairs.

To find the right order in which to make these moves we use Beasley's Second Exit Theorem. Consider the region of Fig. 29(a). The moves we've copied from Fig. 28(b) incorporate just one exit from the region; the vertical jump across P. To make sure there's another we must remove the two pegs on k by a *vertical* pair of to and fro jumps as in Fig. 29(b). But the region of *that* figure can now have only one exit, the vertical jump across f. So our problem's impossible!

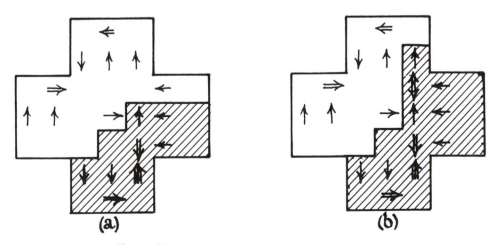

Figure 29. Using Beasley's Second Exit Theorem.

Another Hard Problem

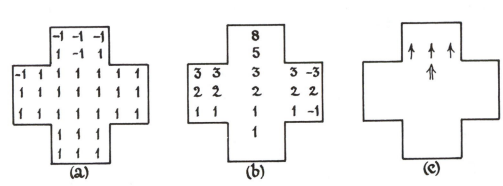

Figure 30. The Reversal Problem *abceg*.

We'll try the 5-peg reversal problem *abceg*, i.e. start with the board full except for spaces at a, b, c, e, g, and finish with pegs in just those places. An equivalent problem is to clear the board of Fig. 30(a), which has negpegs (or **negs**) in each of the places a, b, c, e, g and pegs in the other 28 places.

For the original problem the pagoda function of Fig. 30(b) changes from 20 to 16, or, in the form of Fig. 30(a), from 4 to 0. This pag kills any possible move across b, which would lose 8, so the jumps i_a, k_c shown in Fig. 30(c) are forced, as is the jump j_b to fill b. In order to make this last jump a peg must be delivered to e, and e must also be full by the end, so two jumps x_e are also needed. If we make these five jumps, using negs where needed, we arrive at

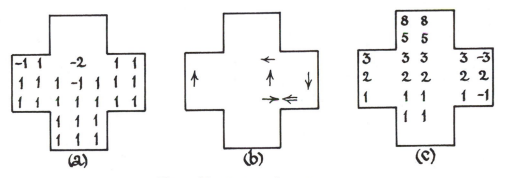

Figure 31. We Make Some Progress.

Fig. 31(a), whose resources are $a^2b^4c^{-1}\alpha\beta$, showing a deficit of a^2c^2. The Deficit Rule tells us we can't make another move of value a^2, since $a^4c^2/c^4 = A^2$, so the jump M_g (Fig. 31(b)) is forced. Now use the pagoda function of Fig. 31(c). Its value for Fig. 31(a) is 2, so the peg at N can't jump inwards, nor can we jump over it upwards, since these moves lose 4 on this peg. So the jump m_G (Fig. 31(b)) is forced. The two pegs at G must now both jump to I, and a peg must be delivered to H for the second of these. This can't come from l, as this loses 4 on the 1st pag, so the jump J_H is also forced. Moreover, as l can't jump downwards, and can't be jumped over (this would lose a)it must make the jump l_j over k. This needs delivery of a peg at k, which can't come from i (loses 6 on the last pag) so the jump I_k is forced. If we make all these moves, which have been collected in Fig. 31(b), we arrive at Fig. 32(a), for which the resources are now

$$a^2 b^3 c^{-1} \alpha = Ab^3 \ \text{ or } \ A.B.b.c\beta \ \text{ or } \ B.a.a\alpha.b\beta.$$

so there is no jump $c\alpha$.

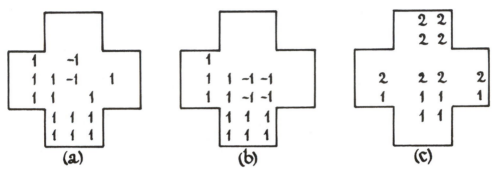

Figure 32. A Cul-de-Sac!

There are two ways in which we might remove the peg at O: by j_{lHJ} or by O_x. The former (after the necessary delivery I_k) leads to Fig. 32(b) which is impossible to clear, as the pag of Fig. 32(c) shows. So O_x is forced (Fig. 33(a)) and this requires the delivery D_P (horizontal delivery is prohibited by the pag of Fig. 31(c)). These two jumps lead to a position whose resources $a^2 \alpha \beta$ are uniquely productive: $(a^2 c^{-1} \alpha)(c\beta)$ and the jump E_x is forced. The L-package of Fig. 33(a) will deliver a second peg to K and the board is cleared by L_J and h_{IJj}.

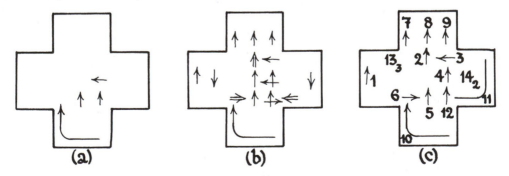

Figure 33. The Problem Solved.

The 23 jumps are shown in Fig. 33(b). How do we do them in practice? In what order? The answer isn't unique, but one possibility is given in Fig. 33(c). It involves two L-packages, 10 and 11, and two chain moves, 13_3 and 14_2.

The Spinner

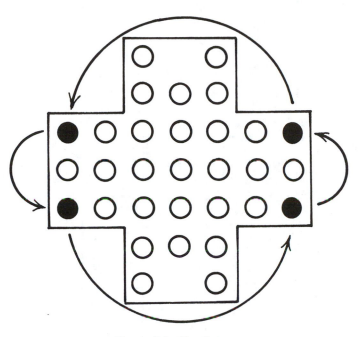

Figure 34. The Spinner.

If you start with empty spaces at b, B and marked pegs at g, M, G, m, can you finish with just the four marked pegs on the board in the respective positions M, G, m, g?

Extras

Our Fine Finalist

The Rule of Two and the Rule of Three together tells us that if the initial hole is at B, then a finalist that starts at J must end in either B or b. Here's a solution for b:

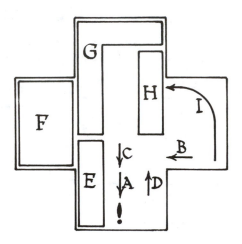

The letters indicate the order of the successive moves, except for the finalizing flourish. The bent arrow we've used for move I is our notation for an L-*package*, as distinct from an L-*purge*.

However, it's impossible for the finalist to finish at B. This is because there are forced moves

$$J_B \quad x_E \quad x_E \quad J_B$$

which consume

$$a^2 \quad c\alpha \quad c\alpha \quad a^2$$

on the Balance Sheet, giving a deficit of a^4c^2. Since

$$\text{deficit}/c^4 = a^4c^{-2}$$

is productive, your remaining resources *aren't*.

Doing the Splits

If you start from Fig. 16(a) and make the moves A to I indicated in Fig. 35, where the pairs of circles C, F, G indicate 2-purges, you'll reach a 5-peg configuration which can easily be reduced to I, L or f. (We found this solution by the ordering process after subtracting this 5-peg configuration from the starting position.)

834

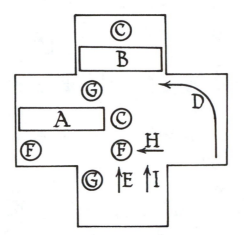

Figure 35. The Train was going to *IvanicGrad*, Ljubljana or *foća*.

All Soluble One-Peg Problems on the Continental Board

All soluble one-peg probems on the Continental board were found by Reiss using his theory. In our language, the Continental board is algebraically equal to its centre, and so for a one-peg problem to be soluble we must have

$$\text{(initial hole)} \times \text{(final peg)} = \text{centre},$$

in our algebraic sense. You can easily check that the initial hole and final peg are at opposite ends of an arrow in

$$(a\,p\,C\,O) \leftrightarrow (A\,P\,c\,o) \quad (e\,G\,J\,M) \leftrightarrow (E\,g\,j\,m).$$

There is a 41-hole board for which Lucas gives all the soluble problems; but see the appendix to his *second* edition, because he first conjectured that most of the problems were insoluble!

The Last Two Moves

The last two moves in Fig. 19(a) are $n_9 G_3$,

A 20-Man Solitaire Army

A 20-man solitaire army can get a scout 4 places out by arranging itself as shown in Fig. 23. The two men with guns can be moved to the shaded places so as to obtain the only other arrangement.

Fool's Solitaire, Etc.

If each of your moves is confined to the middle row or column you'll reach a position like Fig. 36(a) after six jumps. The next pegbound position is the Hammer and Sickle position of Fig. 36(b), reached after ten jumps.

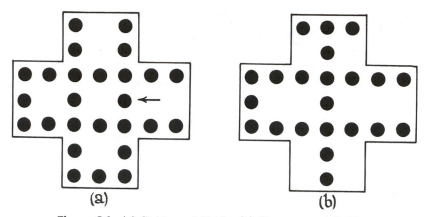

Figure 36. (a) Sickle and Sickle, (b) Hammer and Sickle.

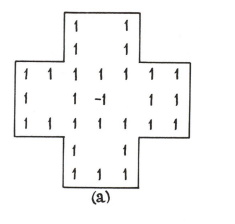

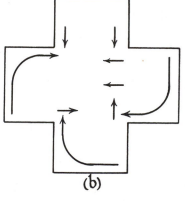

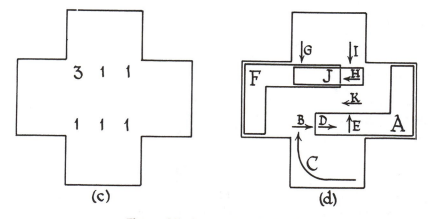

Figure 37. Succouring the Sucker.

To succour the sucker who's made five of the six moves leading to Fig. 36(a) it's best to try to clear Fig. 37(a) to zero. If you set this up on the board (use an upside down peg for the — 1!) you should see how the moves of Fig. 37(b) suggest themselves in order, leading to the easily cleared position, Fig. 37(c). The L-moves in Fig. 37(b) are L-*packages*, not L-purges . You then have the tricky little problem of arranging the moves in order, one solution of which is given in Fig. 37(d), in which A and F are L-*purges*, but C is an L-*package*.

Beasley Proves Bergholt Is Best

Suppose there were a 17-move solution to Central Solitaire. Then Beasley first uses the scoring function of Fig. 38(a) ("score" refers to this function—which is *not* a pagoda function—throughout the proof) and his First Exit Theorem, to show that no move begins or ends on b, n, B or N. The initial and final scores are 20 and 0. Moves which begin or end on b, n, B or N *increase* the score by at least 1. Others decrease it by at most 2 (the careful reader will make a table of score changes for each type of move).

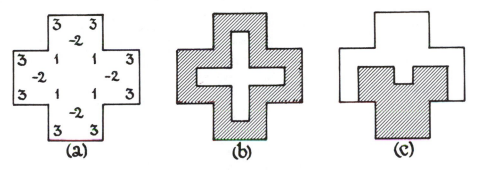

Figure 38. Scoring Function and Regions Used in Beasley's Proof.

Any solution to Central Solitaire contains 11 **reserved** moves:
> the first, which we'll take to be e_x,
> the last, a single jump into the centre,
> the penultimate one, taking a peg to j, p, J or P, and
> eight moves bringing the outside corner pegs to inside corner
> squares so they may be captured.

The first and last moves each increase the score by 2, the penultimate one decreases it by at most 1 and each of the other eight decreases it by at most 2. So the other six (**loose**) moves must decrease the score by at least 7.

The second move is a loose move, either J_j or h_j say. The move J_j doesn't change the score and it leaves the region of Fig. 38(b) full. The first exit from this region is a loose move, either of type b_j or of type (ending with) h_j. The former increases the score by 2 and the other four loose moves would have to decrease it by at least 9, which is impossible. The latter decreases the score by 1, and our four loose moves have to reduce it further by at least 6. If any of these increased the score, the others could not then reduce it to zero, so moves starting or ending on b, n, B or N are again impossible. Such a move *might* occur as the penultimate one, but

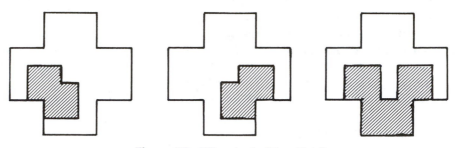

Figure 39. What is the First Exit?

the six loose moves would then have to reduce the score by 10, and the same argument shows this to be impossible.

The second move h_j reduces the score by 1 *and* is a first exit from the region of Fig. 38(b). The other five loose moves must reduce the score by at least 6. What is the first exit from the region of Fig. 38(c)? There are several possibilities, all of them loose moves, which we'll leave the reader to pursue. In some cases he'll want to ask a further question about one of the regions of Fig. 39, whichever is still full. From now on we'll assume that no move begins or ends on b, n, B or N.

How do we clear a, b and c? We've proved that b can't jump out, so there must be a jump over it, say c_a. The two pegs at a now force two jumps a_i and a jump into d, which we shall call a **side delivery**. The four jumps

$$c_a \quad a_i \quad a_i \quad ?_d$$

are parts of at least three moves

$$a_i \ldots \quad ? \ldots_d \quad c_{ai} \ldots \qquad \text{(the normal case), or}$$
$$a_i \ldots \quad ? \ldots_d \quad ? \ldots_{kcai} \qquad \text{(a U-turn).}$$

However a U-turn demands a previous clearance of c and an extra side delivery to f.

Since the same argument applies at n, B and N, we shall need at least four side deliveries, none of which can be among our 11 reserved moves, and none of which can be the first exit from Fig. 38(b). This accounts for 16 moves; call the other the **spare**. Moreover, if a U-turn is involved we have a further side delivery, and so no spare. Note that after eha, p is a side delivery, but j doesn't count as one while g is still occupied, because we'll still need one to clear gnM.

The final stage of Beasley's proof just enumerates all the variations. In the list below the spare move is in **bold**. In the first two variations the first exit from Fig. 38(b) is L, and in all the others it's h. Each variation ends with ‡, § or a colon and a number.

‡   means that the next move can't be a corner move or a side delivery, but the spare has already been used,

§   means that there aren't enough moves left to reduce the score to zero, and

:9   refers to variation number 9, for example.

This list of variations covers the cases where no U-turns are used. If there is a U-turn then is *no* spare move so we have only the variation $ehapc_2$‡(cf. 56).

1	$e\mathbf{J}LCpA_2$‡
2	D§
3	$eh\mathbf{x}af$‡
4	pc_2‡
5	$\mathbf{L}ap$§
6	gj§
7	$\mathbf{k}cPa_2$‡
8	mH§
9	J_2a_2‡
10	G_2‡
11	L_3§
12	mH§
13	J_lG_2‡
14	cP:9
15	L_3G_2ap§
16	c_2§
17	$eh\mathbf{K}MJg_2$‡
18	CD§
19	P_2g_2‡
20	A_2‡
21	d_2g_2‡
22	A_2
23	j_2§
24	CD§
25	P_2A_2‡
26	MJ:19
27	j_2§
28	d_2A_2‡
29	MJ:21
30	$ehacpa$‡
31	$\mathbf{x}$:3
32	$\mathbf{L}$:5
33	$eha\mathbf{j}gpL$§
34	c_2L§
35	J_2M_2‡
36	l_2M_2‡
37	L§
38	$J_2M_2p_dc_2$‡
39	$l_2M_2p_dc_2$‡
40	$ehap\mathbf{c}$:30
41	$\mathbf{k}_2$§
42	$\mathbf{x}$:4
43	$\mathbf{L}$:5
44	$\mathbf{k}mH$§
45	J_lG_2‡
46	L_3§
47	$\mathbf{P}c_2$‡
48	$ehap\mathbf{F}c_3$‡
49	M_2‡
50	c_2MJg_2‡
51	Mc_2:50
52	Jc_3‡
53	g_3‡
54	c_2:50
55	g_2c_2‡
56	c_2P_2‡
57	$\mathbf{F}_2MJg_2$‡
58	$\mathbf{x}$:4
59	$\mathbf{L}$§
60	$\mathbf{j}gL$§
61	J_2M_2‡
62	l_2M_2‡
63	$\mathbf{P}$‡
64	$\mathbf{F}Mj_2$§
65	Jg_2‡

The Classical Problems

These are: start with one empty space, finish with a single peg. They include the reversals, for which Bergholt's results were:

	$a-$	$b-$	$d-$	$e-$	$i-$	$j-$	$x-$	reversal
in	16	18	16	19	16	16	18	moves.

We've just seen that his x-reversal is best possible, but Harry 0. Davis has given a 15-move solution of the i-reversal:

$$kmh_2cPKCD_{PF}A_3MG_2H_4a_2d_5g_3.$$

And here are his solutions, which equal Bergholt's, for the b- and j-reversals:

$$jhapc_2xl_hIf_PA_2GJm_2gL_{Hh}M_2CB_5,$$
$$hKCd_2MJkmH_{Jl}G_3cA_2D_{Fd}g_2ab_7.$$

Hermary identified the 21 distinct problems, one place empty to one place full (see Lucas) and Davis has made a table of best known solutions (see Martin Gardner, "The Unexpected Hanging and Other Mathematical Diversions"). The numbers of moves are:

aa	ap	aO	aC	bb	bn	bx	bB	dd	dK	dH	ee	eM	eJ	ii	il	jj	jg	jE	xx	xb
16	16	17	16	18	17	18	18	16	15	16	19	17	17	15	16	16	16	17	18	17

For this information we thank Wade E. Philpott, who has copies of the solutions. Omar will want to find better ones, or prove them best possible.

References and Further Reading

Martin Aigner, Moving into the desert with Fibonacci, *Math. Mag.*, 70 (1997) 11–21; *MR* 98g:11013.

J.D. Beasley, Some notes on Solitaire, Eureka, 25 (1962) 13-18.

John D. Beasley, *The ins and outs of peg solitaire, Recreations in Mathematics*, 2. Oxford University Press, Eynsham, 1985; *MR* f87c:00002.

John D. Beasley, *The Mathematics of Games, Recreations in Mathematics*, 5, The Clarendon Press, Oxford University Press, New York, 1989; *MR* 90k:90181.

E. Bergholt, The Queen, May 11, 1912; and "The Game of Solitaire", Routledge, London, 1920.

N.G. de Bruijn, A Solitaire game and its relation to a finite field, J. Recreational Math. 5 (1972) 133-137.

Busschop, "Recherches sur le jeu de Solitaire" Bruges, 1879.

M. Charosh, The Math. Student J., U.S.A, March 1961.

Donald C. Cross, Square Solitaire and variations, J. Recreational Math. 1 (1968) 121-123.

Béla Csákány and Rozália Juhász, The solitaire army reinspected, *Math. Mag.*, 73 (2000) 354–362; *MR* 2002b:05012.

Harry O. Davis, 33-solitaire, new limits, small and large. Math. Gaz. 51 (1967) 91-100.

Antoine Deza & Shmuel Onn, Solitaire lattices, *Graphs Combin.*, 18 (2002) 227–243; *MR* 2003d:05021.

H.E. Dudeney, The Strand Magazine, April, 1908; and see "Amusements in Mathematics", problems 227, 359, 360, Nelson, London 1917. pp. 63-64. 107-108. 195. 234.

M. Gardner, Sci. Amer. 206 #6 (June 1962) 156-166; 214 #2 (Feb. 1966) 112-113: 214 #5 (May 1966) 127.

M. Gardner, "The Unexpected Hanging and other Mathematical Diversions", Simon and Schuster 1969, p.126.

Miguel de Guzmàn, *The Countingbury Tales, Fun with Mathematics*, Translated by Jody Doran. World Scientific Publishing Co., Inc., River Edge NJ, 2000; *MR* 2002g]:00002.

Heinz Haber, Das Solitaire-Spiel, in "Das Mathematische Kabinett", Vol. 2, Bild der Wissenschaft, D. V-A., Stuttgart 1970, pp. 53-57.

Irvin Roy Hentzel, Triangular puzzle peg, J. Recreational Math. 6 (1973) 280-283.

Hermary, Sur le jeu du Solitaire, Assoc. franc, pour l'avancement des sci., Congres de Montpellier, 1879.

Ross Honsberger, "Mathematical Gems II", Mathematical Association America, 1976, chap. 3, 23-28.

Masashi Kiyomi and Tomomi Matsui, Integer programming based algorithms for peg solitaire problems. Algorithm engineering as a new paradigm (Japanese) (Kyoto, 2000), Sūrikaisekikenkyūsho Kōkyūroku No. 1185 (2001), 100–108.

M. Kraitchik, "Mathematical Recreations", George Alien & Unwin, London, 1943, pp. 297-298 quotes letter 1716:1:17 Leibniz to Monmort.

G. W. Leibniz, L'estime des apparences, 21 manuscrits de Leibniz sur les probabilités, la théorie des jeux, l'espérance de vie, Translated, annotated and with an introduction by Marc Parmentier, *Mathesis*, Librairie Philosophique J. Vrin, Paris, 1995; *MR* 96m:01009.

E. Lucas, "Récréations Mathématiques", Blanchard, Paris 1882, Vol. 1, part 5, 89-141 is mainly concerned with the continental board (37 places) but pp. 132-138 refer to the English board and much of the whole is applicable. He attributes (pp. 114-115) the 3-*purge* to Hermary.

Cristopher Moore and David Eppstein, One-dimensional peg solitaire, and duotaire, in *More Games of No Chance* (Berkeley CA, 2000), 341–350, *Math. Sci. Res. Inst. Publ.*, 42, Cambridge Univ. Press, Cambridge, 2002.

M. Reiss, Beitrage zur Theorie der Solitär-Spiels, Crelle's J., 54 (1857) 344-379.

Ruchonet, Théorie du solitaire, par feu le docteur Reiss, librement traduit de l'allemand, Nouv. Corr. math., t. Ill p. 231, Bruxelles 1877.

B.M. Stewart, "Theory of Numbers," Macmillan, New York, 1952, 1964, pp. 20-26. Analyzes Solitaire on a 7 × 5 rectangular board. He colors the diagonals with 3 colors in either direction and obtains the Rule of Three as an example of congruences (mod 3). Exercise 4.5 on p. 24, due to F. Gozreh, asks you to start from an even length row of pegs with the second peg missing and finish with just the second peg from the other end. A pattern of 2-purges does the trick. If you start with the fifth peg missing, then a 2-purge and a double jump by the 1st peg reduces the problem to the earlier one. Can you clear the row with other pegs missing? Which missing pegs enable you to clear an *odd* length row?

-24-

Pursuing Puzzles Purposefully

> The chapter of accidents is the longest chapter in the book.
> John Wilkes

> I shall proceed to such Recreations as adorn the Mind;
> of which those of the Mathematicks are inferior to none.
> William Leybourne; *Pleasure with Profit.*

We know you want to use your winning ways mostly when playing with other people, but there are quite a lot of puzzles that are so interesting that you really feel you're playing a game against some invisible opponent—perhaps the puzzle's designer—maybe a malevolent deity. In this chapter we'll discuss a few cases where some kind of strategic thinking simplifies the problem. But because we don't want to spoil your fun we'll try to arrange not always to give the *whole* game away.

Soma

This elegant little puzzle was devised by Piet Hein. Figure 1 shows the seven non-convex shapes that can be made by sticking 4 or fewer $1 \times 1 \times 1$ cubes together. Piet Hein's puzzle is to assemble these as a $3 \times 3 \times 3$ cube.

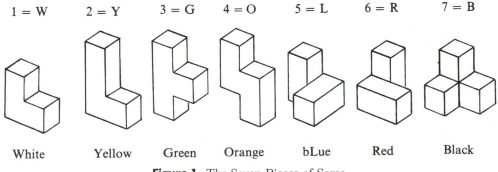

1 = W	2 = Y	3 = G	4 = O	5 = L	6 = R	7 = B
White	Yellow	Green	Orange	bLue	Red	Black

Figure 1. The Seven Pieces of Soma.

We advise you to use seven different colors for your pieces as in the figure. Many people solve this puzzle in under ten minutes, so it can't be terribly hard. But we've got a distinct feeling that it's much harder than it ought to be. Is this just because the pieces have such awkwardly wriggly shapes?

Blocks-in-a-Box

Here is another puzzle invented by one of us some years ago, in which all the pieces are rectangular cuboids but it still seems undeservedly hard to fit them together. We are asked to pack one $2 \times 2 \times 2$ *cube*, one $2 \times 2 \times 1$ *square*, three $3 \times 1 \times 1$ *rods* and thirteen $4 \times 2 \times 1$ *planks* into a $5 \times 5 \times 5$ box (Fig. 2). It's quite easy to get all but one of the blocks into the box, but somehow one piece always seems to stick out somewhere. A friend of ours once spent many evenings without ever finding a solution. Why is it so much harder than it seems to be?

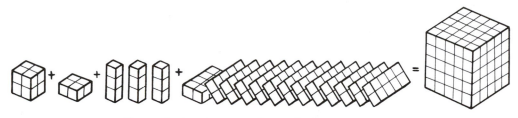

Figure 2. The Eighteen Pieces for Blocks-in-a-Box.

Hidden Secrets

In our view the good puzzles are those with simple pieces but difficult solutions. Anyone can make a hard puzzle with lots of complicated pieces but how can you possibly make a hard puzzle out of a few easy pieces?

When a seemingly simple puzzle is unexpectedly difficult, it's usually because, as well as the obvious problem, there are some hidden ones to be attended to. Both Soma and Blocks-in-a-Box have such hidden secrets, but let's look at a much simpler puzzle, to fit six $2 \times 2 \times 1$ squares into a $3 \times 3 \times 3$ box, leaving three of the $1 \times 1 \times 1$ cells empty—the *holes* (Fig. 3). This now seems fairly trivial, but even so there's a hidden secret which sometimes makes people take more than 5 minutes over it. This hidden problem comes from the fact

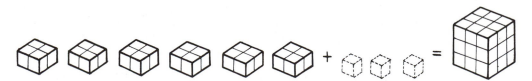

Figure 3. A Much Simpler Puzzle.

that the square pieces can only occupy an even number of the cells in each horizontal layer. So since 9 is odd each horizontal layer must have a hole and there are only just enough holes to go round. Of course these holes must also manage to meet each of the three layers in each of two vertical directions—you can't afford to have two holes in any layer, because some other layers would have to go without.

So the problem wasn't really to fit the *pieces* in but rather the *holes*. Only when you've realized this do you see why the unique solution (Fig. 4) has to be so awkward looking, with the holes strung out in a line between opposite corners rather than neatly arranged at the top of the box.

Perhaps you'd like to try the big Blocks-in-a-Box problem now, before looking at the extra hints in the Extras.

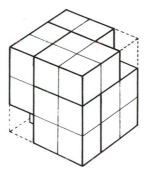

Figure 4. Six Squares in a 3 × 3 × 3 Box.

The Hidden Secrets of Soma

It's because the Soma puzzle pieces have to satisfy some hidden constraints as well as the obvious ones, that it causes most people more trouble than it should. Let's see why.

The 3 × 3 × 3 cube has 8 *vertex* cells, 12 *edge* cells, 6 *face* cells and 1 *central* cell as in Fig. 5.

Figure 5. The Vertex, Edge, Face and Central (invisible) Cells.

Now the respective pieces can occupy at most

$$\begin{array}{ccccccc} W & Y & G & O & L & R & B \\ 1, & 2, & 2, & 1, & 1, & 1, & 1 \end{array}$$

of the vertex cells, so just one piece, the **deficient** one, must occupy just one less vertex-cell than it might. The green piece can't be deficient without being doubly so, and therefore:

> the Green piece has
> its spine along an
> edge of the cube.

Now let's color the 27 cells of the cube in two alternating colors,

Flame for the 14 FaVored cells, F and V,
Emerald for the 13 ExCeeded ones, E and C.

Then in *one* solution that we know, the respective pieces occupy

$$\begin{array}{ccccccc} W & Y & G & O & L & R & B \\ 2 & +2 & +3 & +2 & +2 & +2 & +1 = 14 \text{ F, V cells,} \\ 1 & +2 & +1 & +2 & +2 & +2 & +3 = 13 \text{ E, C cells,} \end{array}$$

but the Yellow, Orange, bLue and Red pieces, and we now know also the Green piece, *must* occupy these numbers in *every* solution, and therefore so must the White and the Black, since an interchange of colors in either or both of these would alter the totals.

> The White piece occupies
> 2 FV cells, 1 EC cell.

> The Black piece occupies
> 1 FV cell and 3 EC ones.

For the placing of a single piece within the box, these considerations leave only the positions of Fig. 6 (which all arise). You'll see that up to symmetries of the cube, the placement of any single piece is determined by whether or not it is deficient and whether or not it occupies the central cell.

The hidden secrets of Soma make it quite likely that one of the first few pieces you put in may already be wrong, when of course you'll spend a lot of time assembling more pieces before such a mistake shows its effect. This would happen for instance if you started by putting the

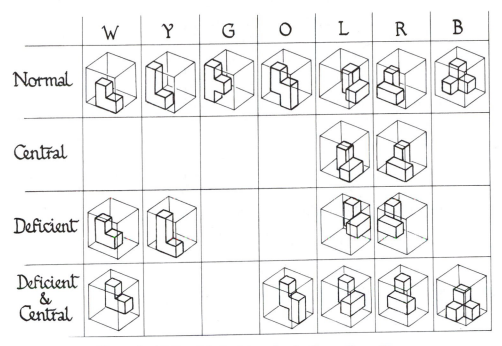

	W	Y	G	O	L	R	B
Normal							
Central							
Deficient							
Deficient & Central							

Figure 6. All Possible Positions for the Seven Soma Pieces.

corner of the White piece into a corner of the cube. But if you only put the pieces into the allowed positions, you'll find a solution almost as soon as you start. The complete list of 240 Soma solutions was made by hand by J.H. Conway and M.J.T. Guy one particularly rainy afternoon in 1961. The SOMAP in the Extras enables you to get to 239 of them, when you've found one—*and* located it on the map!

Hoffman's Arithmetico-Geometric Puzzle

A well-known mathematical theorem is the inequality between the arithmetic and geometric means:

$$\sqrt{ab} \leq \frac{a+b}{2}.$$

Figure 7 provides a neat proof of this in the form

$$4ab \leq (a+b)^2$$

and the three variable version

$$27abc \leq (a+b+c)^3$$

has prompted Dean Hoffman to enquire whether 27 $a \times b \times c$ blocks can always be fitted into a cube of side $a + b + c$. This turns out to be quite a hard puzzle if a, b, c are fairly close

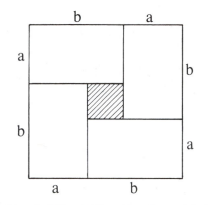

Figure 7. Proof of the Arithmetico-Geometric Inequality.

together but not equal. A good practical problem is to fit

$$27 \ 4 \times 5 \times 6 \text{ blocks into a } 15 \times 15 \times 15 \text{ box.}$$

With these choices, as for any others with

$$\frac{1}{4}(a + b + c) < a < b < c,$$

it can be shown that each vertical stack of three blocks must contain just one of each height a, b, c, while there must be just three of each height in each horizontal layer. There must be the same unused area on each face (just 3 square units in the $4 \times 5 \times 6$ case).

It's almost impossible to solve the puzzle if you don't keep these hidden secrets constantly in mind because you'll make irretrievable mistakes like making a stack of three height 5 blocks, or leaving a 2×2 empty hole on some face. When you *do* keep them in mind, the puzzle becomes much easier, being only extremely difficult! You'll find some information about solutions to Hoffman's puzzle in the Extras.

Coloring Three-by-Threee-by-Three by Three, Bar Three

In Hoffman's $3 \times 3 \times 3$ puzzle, the three lengths along any line of three had to be different. Can you color the cells of a $3 \times 3 \times 3$ tic-tac-toe board with

three different colors,

using all

three colors the same

number (9) of times, in such a way that *none* of the $\frac{1}{2}(5^3 - 3^3) = 49$ tic-tac-toe lines uses

three different colors,

nor has all its

three colors the same?

Wire and String Puzzles

Figure 9 shows a number of topological puzzles which can be made with wire and string. It's a pity that manufacturers don't seem to know about all of these.

You wouldn't expect to be able to say much about such varied looking objects, but in fact there's a quite general principle which helps you to solve a lot of them.

The Magic Mirror Method

We'll just take the one-knot version of the puzzle shown in Fig. 9(c) which has been commercially sold as The Loony Loop (Trolbourne Ltd., London). You're to take the string off the rigid wire frame in Fig. 8(a).

If only that rigid wire were a bit stretchable, the puzzle would be quite easy. After squashing the string up (Fig. 8(b)) so as not to get in the way, we could stretch the loops over the ends (Fig. 8(c)) and shrink them again (Fig. 8(d)). After this we can take the string right off (Fig. 8(e)) and then put the loops back as they were (Fig. 8(f)) so as not to upset the owner.

Now the change from Fig. 8(b) to Fig. 8(d) could be accomplished by continuously distorting space. Think of embedding the puzzle in a flexible jelly, if Mother has one made up. Now old-fashioned fairgrounds had special mirrors which seemed to distort space in very funny ways. Now let's imagine a magic mirror with the wonderful property that the distortion is just what's required to make Fig. 8(b) look like Fig. 8(d). Hold the wire frame absolutely still before the magic mirror (Fig. 10(a)) and bunch the string up until it's almost a single point on the axis. Because the space distortion was continuous, its image will also be almost a single point on the image axis.

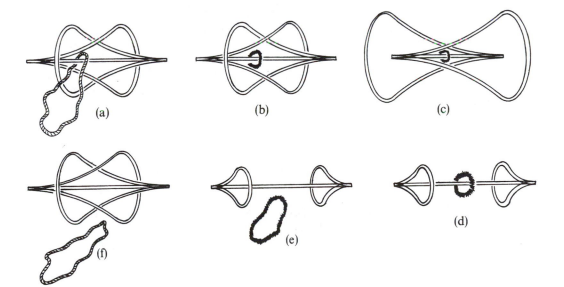

Figure 8. Solving The Loony Loop.

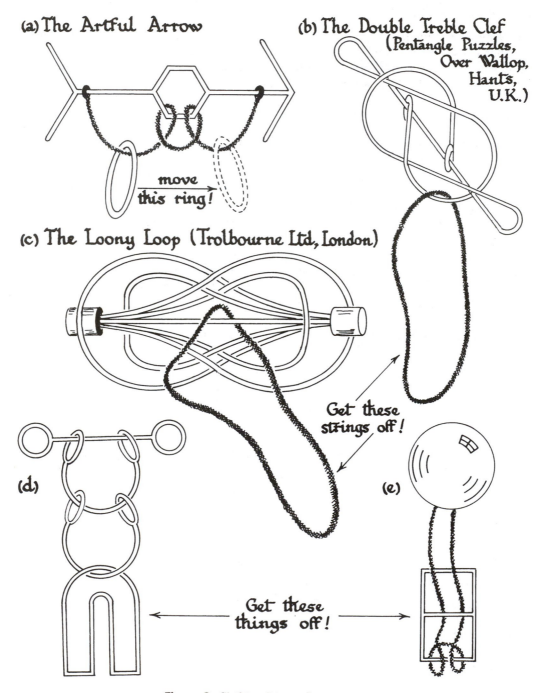

Figure 9. Shifting Rings, Strings . . . . . .

(f) Ball and Chain

(g)

(h)

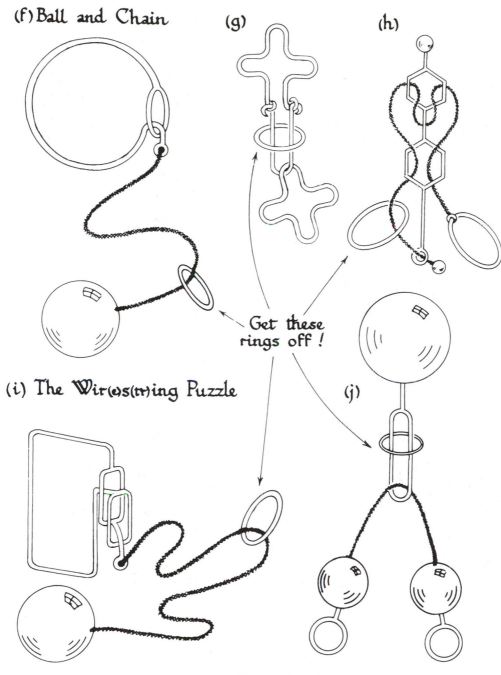

Get these
rings off !

(i) The Wir(e)s(tr)ing Puzzle

(j)

...... and Other Things.

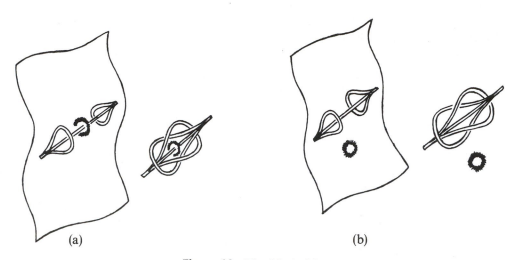

(a)   (b)

Figure 10. The Magic Mirror.

Now, very carefully, move the string in just such a way that its *image* in the magic mirror moves completely away from the wire and shrinks to a small point at some little distance from it. Once again, because the distortion was continuous, the real string must now be almost a point, some distance from the wire, and you've solved the puzzle. Easy, wasn't it?

In such cases it often helps to imagine an intermediate distortion. In Fig. 11 we show two stages in an intermediately distorted one-knot Loony Loop . Perhaps you're ready for the two knot version (Fig. 9(c))? Or the Double Treble Clef (Fig. 9(b)) (Pentangle Puzzles, Over Wallop, Hants, U.K.)?

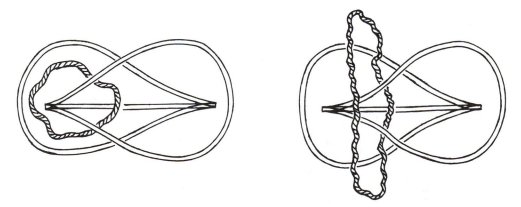

Figure 11. A Less Distorting Mirror.

If a puzzle has got just one completely rigid piece and a number of completely flexible pieces then you can often use the magic mirror method to pretend that the rigid piece is also flexible. For instance, although it may seem impossible to make the braided piece of paper in

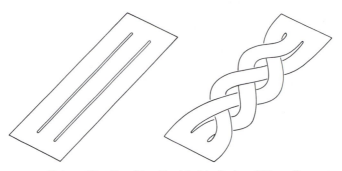

Figure 12. Can You Braid this Strip of Paper?

Fig. 12 without glue, it can be undone quite easily. This principle is quite familiar to craftsmen in leather. (To make it you should start braiding at one end and undo the tangle which forms at the other.)

The Barmy Braid

The Barmy Braid problem appears for the first time in this book. It's to take the string off the rigid wire frame in Fig. 13(a). You know you can do it, because in a suitable magic mirror it looks like Fig. 13(b).

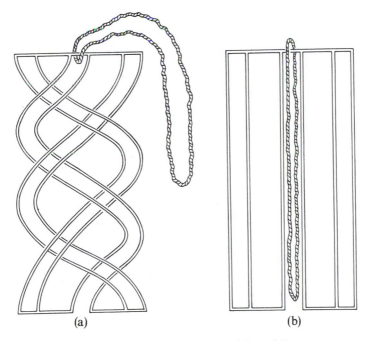

(a)   (b)

Figure 13. Barmy Braid Meets Magic Mirror.

The Artful Arrow

Figure 8(a) is our version of a puzzle that appears in many different forms. The basic framework is often a bar of wood with a drill hole in place of our hexagon. We have even seen a version in which the ends of our arrow are a giant's arms and the central hole his nostrils, but the solution is always the same! You can solve this puzzle, and some similar ones, by a modification of the Magic Mirror Method which we call

The Magic Movie Method

If the Artful Arrow had a much smaller ring, there'd be no difficulty about solving it; we'd just slide the ring along the string from the tail of the arrow to its head. Let's suppose we have a kinematic friend who takes a movie of this, but that through some accident with his filters, the string doesn't show up too well, so that what the movie shows is the rigid arrow framework and a little ring that wanders about in space. In fact the ring moves downward through the hexagon (1 to 2 to 3 in Fig. 14(a)), sweeps around (3 to 4 to 5) and then comes safely back up again (5 to 6 to 7).

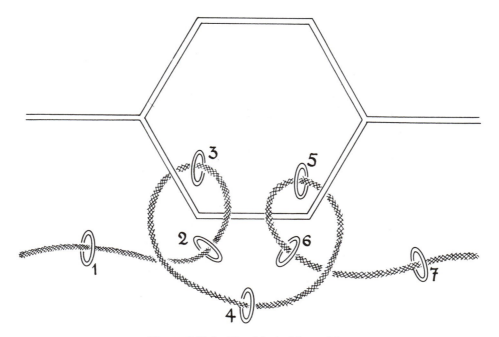

Figure 14(a). The Magic Movie M_0.

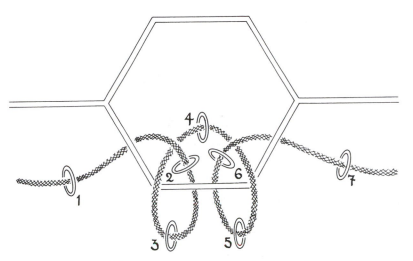

Figure 14(b). An Intermediate Half-Magic Movie.

What we want to do is to watch this movie in a sort of hyperspace magic mirror which distorts both space and time. Our friend can arrange this for us by taking the movie M_0 to the animation department where they can change the whole movie bit by bit, first to M_1, in which the ring goes down through the hexagon and wanders about a bit less before it comes back up again, then to M_2, in which it wanders hardly at all before coming up, then to M_3 in which it only takes a timid dip through the hexagon, then in M_4 not at all, while in M_5, M_6 ...the size of the ring gradually increases until it is too big to go through the hexagon.

The trouble with all these movies is that we can't see the string! But since we intend the sequence of movies to realize a continuous distortion of space-time, we can ask the animation department to work overtime and fill in the position of the string as well. The final movie, M_{10} say, should satisfy the producer as representing a solution to the puzzle.

As usual, it helps if the whole process is only half-magic. What must actually happen in this sequence of movies is that the excursion of the ring through the hexagon is gradually replaced by a pulling up of the central loop of string (Mahomet coming to the mountain). In Fig. 14(b) we show an intermediate movie in which you can hardly tell whether this loop, as it passes through the ring in position 4, is above or below the hexagon. You can therefore solve this puzzle by passing the ring from 1 to 2 to 3 while the loop is *below* the hexagon, then lifting the loop a bit while you slide the ring from 3 to 4 to 5 and drop it again so that you can go from 5 to 6 to 7. Since all these movies can be made with a full-sized ring, this will solve the puzzle.

This argument allows us to extend the idea we noted when introducing the Barmy Braid. Suppose that a puzzle has *any number* of rigid pieces (like our arrow and ring) and some arbitrarily flexible ones (e.g., our string) and you could find a solution if the rigid pieces were made flexible. Then, if the motion of the rigid pieces in your solution can be continuously distorted into a rigidly permissible motion, you can use the Magic Movie Method to solve the original puzzle. In topologists' technical language we are using the *Isotopy Extension Principle*.

Party Tricks and Chinese Rings

Figure 15. Girl Meets Boy.

You must have met the party trick where the boy and the girl have to separate themselves without untying the knots in the string. Usually they have lots of fun stepping through one another's arms without effect before they find the real answer.

Let's look at one of those fists more closely (Fig. 16(a)). With a really magic mirror this looks like (Fig. 16(b)) and the solution is obvious, but as usual it's slightly easier to see what to do if your mirror is only half magic (Fig. 16(c)).

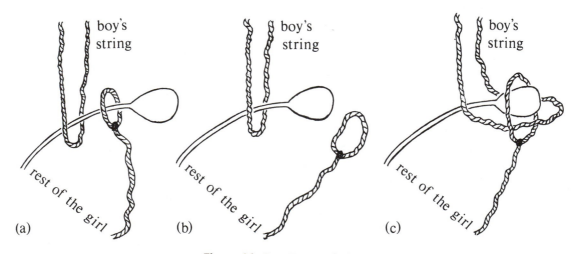

Figure 16. Boy Leaves Girl.

Figure 17(a). Pajamas on Hanger.

One of our wire and string puzzles is very like this. The pajama-shaped frame at the bottom of Fig. 17(a) is made of wire rather than string, but it happens to be just about the shape that a piece of string would need to get to while being taken off. In Fig. 8(d) you'll see there's a similar puzzle, but with an extra piece.

The magic mirror in Fig. 17(b) shows that this puzzle can certainly be solved if the wire pajama shape is replaced by a completely flexible string—once again this funny shape is sufficient to overcome its lack of flexibility.

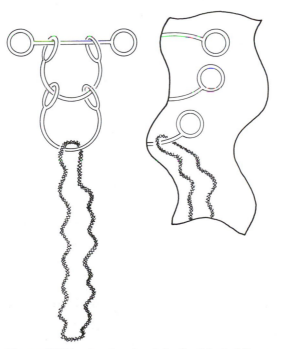

Figure 17(b). Another Look in the Magic Mirror.

The Chinese rings are an indefinite extension of this principle. The magic mirror method shows that the string in Fig. 18(a) can be taken right off. In the course of doing so it reaches a position like that of the wire loop in Fig. 18(b), and removal of this is the usual Chinese Rings puzzle.

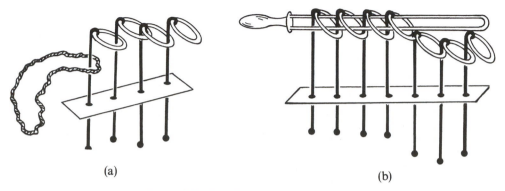

(a)   (b)

Figure 18. The Chinese (st)ring Puzzle.

Chinese Rings and the Gray Code

Figure 19(a) shows a certain position of a 7-ring Chinese Rings puzzle. We call this position

$$1 \quad 0 \quad 1 \quad 1 \quad 1 \quad 0 \quad 0$$

because the rings we've numbered

$$64 \quad 32 \quad 16 \quad 8 \quad 4 \quad 2 \quad 1$$

are respectively

$$\text{on} \quad \text{off} \quad \text{on} \quad \text{on} \quad \text{on} \quad \text{off} \quad \text{off}$$

the loop. ("On" means that the ring's retaining wire passes through the loop.) Which positions neighbor this?

You hardly need a magic mirror to see how the state of the rightmost ring, number 1, can always be changed (Fig. 19(b)), showing that our position neighbors

$$1 \quad 0 \quad 1 \quad 1 \quad 1 \quad 0 \quad 1.$$

But it also neighbors

$$1 \quad 0 \quad 1 \quad 0 \quad 1 \quad 0 \quad 0$$

as well!

To see this, slip ring number 8 up over the end of the loop as suggested by the dotted arrow in Fig. 19(a) and then drop it down through the loop as hinted in Fig. 19(c).

In general the rightmost ring, number 1, can always be slipped on or off the loop, so that

$$\ldots \ ? \ ? \ ? \ 0 \quad \text{neighbors} \quad \ldots \ ? \ ? \ ? \ 1.$$

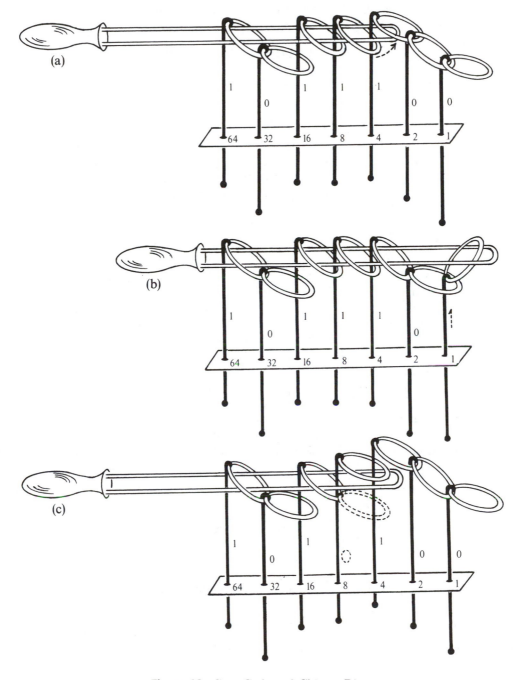

Figure 19. Gray Code and Chinese Rings.

But also a ring can be slipped on or off provided that the ring just right of it is *on* and all ring right of *that* are *off*, so that

$$\ldots \ ?\ 1\ 1\ 0\ 0\ 0\ \text{ neighbors }\ \ldots\ ?\ 0\ 1\ 0\ 0\ 0.$$

With these neighboring rules the entire set of 2^n positions in the n-ring puzzle form one continuous sequence, which for $n = 4$ is:

	ring #	8	4	2	1	
state #	8, i.e.	1	0	0	0	is 15 moves from being off,
state #	9, i.e.	1	0	0	1	is 14 moves from being off,
state #	11, i.e.	1	0	1	1	is 13 moves from being off,
state #	10, i.e.	1	0	1	0	is 12 moves from being off,
state #	14, i.e.	1	1	1	0	is 11 moves from being off,
state #	15, i.e.	1	1	1	1	is 10 moves from being off,
state #	13, i.e.	1	1	0	1	is 9 moves from being off,
state #	12, i.e.	1	1	0	0	is 8 moves from being off,
state #	4, i.e.	0	1	0	0	is 7 moves from being off,
state #	5, i.e.	0	1	0	1	is 6 moves from being off,
state #	7, i.e.	0	1	1	1	is 5 moves from being off,
state #	6, i.e.	0	1	1	0	is 4 moves from being off,
state #	2, i.e.	0	0	1	0	is 3 moves from being off,
state #	3, i.e.	0	0	1	1	is 2 moves from being off,
state #	1, i.e.	0	0	0	1	is 1 moves from being off,
and state #	0, i.e.	0	0	0	0	is OFF!

How do we tell how many moves it takes to get all the rings off if we're given only the state number, i.e. the sum of the numbers of the rings that are on? The answer displays a remarkable connexion with nim-addition! When you're in state number n, it will take you exactly

$$n \overset{*}{+} \lfloor n/2 \rfloor \overset{*}{+} \lfloor n/4 \rfloor \overset{*}{+} \lfloor n/8 \rfloor \overset{*}{+} \ldots = m$$

moves to get off. For example in state 13 you're just

$$13 \overset{*}{+} 6 \overset{*}{+} 3 \overset{*}{+} 1 = 9$$

moves away. And if you're given a number m, then state number

$$m \overset{*}{+} \lfloor m/2 \rfloor = n$$

is the one that's just m moves from off. For example

$$9 \overset{*}{+} 4 = 13.$$

Let's find the position that's 99 moves from off in the 7-ring puzzle. Because the binary expansions of 99 and $\lfloor 99/2 \rfloor$ are

$$
\begin{array}{cccccccc}
& 1 & 1 & 0 & 0 & 0 & 1 & 1 \\
\text{and} & & 1 & 1 & 0 & 0 & 0 & 1 \\
\hline
\text{the answer is} & 1 & 0 & 1 & 0 & 0 & 1 & 0.
\end{array}
$$

How many moves is state

$$
\begin{array}{ccccccc}
1 & 1 & 0 & 1 & 1 & 1 & 1
\end{array}
$$

from off? The answer is found by the 7-term nim-sum

$$
\begin{array}{ccccccc}
1 & 1 & 0 & 1 & 1 & 1 & 1 \\
  & 1 & 1 & 0 & 1 & 1 & 1 \\
  &   & 1 & 1 & 0 & 1 & 1 \\
  &   &   & 1 & 1 & 0 & 1 \\
  &   &   &   & 1 & 1 & 0 \\
  &   &   &   &   & 1 & 1 \\
  &   &   &   &   &   & 1 \\
\hline
= & 1 & 0 & 0 & 1 & 0 & 1 & 0,
\end{array}
$$

which is the binary expansion of 74.

In various kinds of control device it's important to code numbers in such a way that the codes from adjacent numbers differ in only one place and the code that appears above, known to engineers as the Gray code, has this useful property. It has also been used in transmitting television signals. However, its connexion with the Chinese Rings puzzle was known to Monsieur L. Gros, more than a century ago. Incidentally, the multiknot Loony Loop is connected with a ternary version of the Gray code.

The Chinese Rings have occasionally been used as a sort of combination lock. In recent years several mechanical and electronic puzzles, completely different in appearance, but employing the same mathematical structure, have appeared on the market.

The Tower of Hanoï

In happier times, Hanoï was mainly known to puzzlers as the fabled site of that temple where monks were ceaselessly engaged in transferring 64 gold discs from the first to the last of three pegs according to the conditions that

only one disc may be moved at a time, and
no disc may be placed above a smaller one.

Figure 20(a) shows the initial position in a smaller version of the puzzle and Fig. 20(b) shows the position 13 moves later.

In this puzzle it's possible to make mistakes, unlike in the Chinese Rings where the only mistake you can make is to start travelling in the wrong direction. However, you won't make too many mistakes if you use discs that are alternately gold and silver and

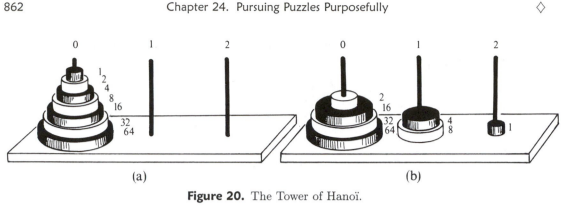

Figure 20. The Tower of Hanoï.

> never place a disc immediately
> above another of the same metal.

To find out where you should be after m moves, expand m in binary, and then, according as the total number of discs is

	even	or	odd,
replace a 1 digit by the ternary number	1	or	2,
replace a 2 digit by the ternary number	21	or	12,
replace a 4 digit by the ternary number	122	or	211,
replace an 8 digit by the ternary number	2111	or	1222,
replace a 16 digit by the ternary number	12222	or	21111,
replace a 32 digit by the ternary number	211111	or	122222,
replace a 64 digit by the ternary number	1222222	or	2111111,
....	....	...	....

These ternary numbers, when added mod 3 without carrying, show you what peg each disc should be on. For 13 moves and a 7-disc tower, since 7 is odd and

$$\left. \begin{array}{r} 1 \\ +4 \\ +8 \\ \hline = 13 \end{array} \right\} \quad \text{we find the ternary numbers} \quad \left\{ \begin{array}{r} 2 \\ 211 \\ 1222 \\ \hline 0001102, \end{array} \right.$$

showing that disc 1 should be on peg 2, discs 4 and 8 on peg 1, and the rest on peg 0 as in Fig. 20(b).

The Tower of Hanoï puzzle and the fable which usually accompanies it were invented by Messieurs Claus (Édouard Lucas) and De Parville in 1883 and 1884.

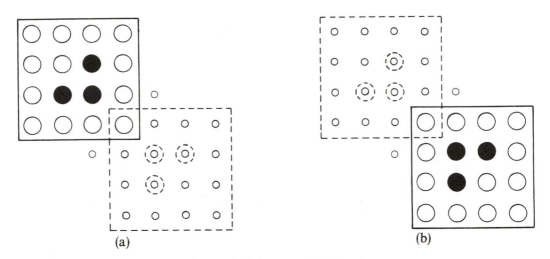

Figure 21. A Solitaire-Like Puzzle.

A Solitaire-Like Puzzle and Some Coin-Sliding Problems

A little puzzle we came across recently is played in a way very similar to the game of Peg Solitaire, except that the pegs are not removed after jumping. Starting from the position of Fig. 21(a), go to the "opposite" position of Fig. 21(b) jumping only in the N-S and W-E directions. The three special pegs are to move to the three special places.

H	H	H		T	T	T

Figure 22. Swap the Hares and Tortoises.

This is rather like various two-dimensional forms of the familiar Hares and Tortoises (or sheep and goats) puzzle (Fig. 22) in which the animals (you can use coins) have to change places and the permitted moves are as in the game of Toads and Frogs in Chapter 1. Other problems with the same coins are:

1. get from Fig. 23(a) to Fig. 23(b) with just 3 moves of 2 contiguous coins (the coins to be slid on the table, remaining in the same orientation and touching throughout);

2. the same, but reversing the orientation of each pair of coins as it is moved;

3. similar problems, but with more coins;

4. form the six coins of Fig. 24(a) into a ring (Fig. 24(b)) with just three moves. At each move one coin must be slid on the table, without disturbing any of the others, and positioned by touching it against just two coins. For example, you might try Fig. 24(c) for your first move, but then you wouldn't be able to slide the middle one out.

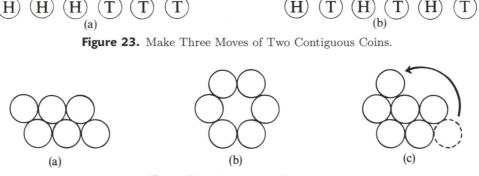

Figure 23. Make Three Moves of Two Contiguous Coins.

Figure 24. Ringing the Changes.

The Fifteen Puzzle and the Lucky Seven Puzzle

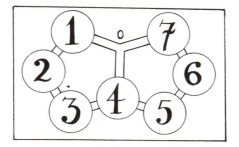

Figure 25. Sam Loyd's Fifteen Puzzle.

The most famous sliding puzzle is Sam Loyd's *Fifteen Puzzle* in which the home position is Fig. 25 and the move is to slide one square at a time into the empty space. You are required to get home from the random position you usually find the puzzle in. Nowadays the puzzle is usually sold with pieces so designed that it is impossible to remove them from the base.

Figure 26. The Lucky Seven Puzzle.

A more interesting puzzle is the **Lucky Seven Puzzle**, for which the home state is displayed in Fig. 26 and similar rules apply.

In such puzzles there are certain basic permutations of the pieces that bring the empty space back to its standard position. For the Seven Puzzle you can either move the four discs in the left pentagon in the order 1, 2, 3, 4, 1, leading to the position of Fig. 27(a) or treat the right pentagon similarly, moving 7, 6, 5, 4, 7, leading to the position of Fig. 27(b).

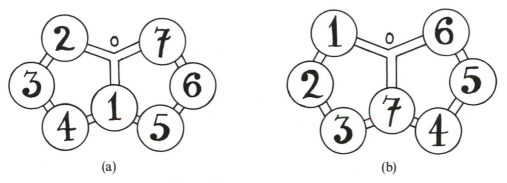

(a)   (b)

Figure 27. After a Few Moves.

In the first case we have effected the permutation α in which

$$\left.\begin{array}{rccccccc} \text{disc} & 1 & 2 & 3 & 4 & 5 & 6 & 7 \\ \text{goes to place} & 4 & 1 & 2 & 3 & 5 & 6 & 7 \end{array}\right\} \begin{array}{l}\text{or, for short,} \\ (1432)\,(5)\,(6)\,(7),\end{array}$$

and in the second case the permutation β in which

$$\left.\begin{array}{rccccccc} \text{disc} & 1 & 2 & 3 & 4 & 5 & 6 & 7 \\ \text{goes to place} & 1 & 2 & 3 & 5 & 6 & 7 & 4 \end{array}\right\} \text{or } (1)\,(2)\,(3)\,(4567).$$

We can obviously combine these basic permutations to any extent. For instance, by performing the sequence

$$\begin{array}{ccccccccc}
1 & \xrightarrow{\alpha} 4 & \xrightarrow{\beta} 5 & \xrightarrow{\alpha} 5 & \xrightarrow{\alpha} 5 & \xrightarrow{\beta} 6 \\
2 & \to 1 & \to 1 & \to 4 & \to 3 & \to 3 \\
3 & \to 2 & \to 2 & \to 1 & \to 4 & \to 5 \\
4 & \to 3 & \to 3 & \to 2 & \to 1 & \to 1 \\
5 & \to 5 & \to 6 & \to 6 & \to 6 & \to 7 \\
6 & \to 6 & \to 7 & \to 7 & \to 7 & \to 4 \\
7 & \to 7 & \to 4 & \to 3 & \to 2 & \to 2
\end{array}$$

$$\left.\begin{array}{rccccccc} \text{disc} & 1 & 2 & 3 & 4 & 5 & 6 & 7 \\ \text{goes to place} & 6 & 3 & 5 & 1 & 7 & 4 & 2 \end{array}\right\} \text{or } (164)\,(2357).$$

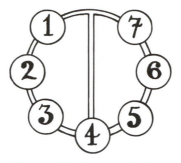

Figure 28. Crossing Bridges.

By combining any given permutations in all possible ways we get what mathematicians call a **group** of permutations. Is there an easy way to see which permutations belong to the group of the Lucky Seven Puzzle? Yes! The trick, as always in such cases, is to find some permutations which keep most of the objects fixed. In the case of the Seven Puzzle it seems best to regard the outer edges as forming a complete circle across which there is a single bridge between places 0 and 4 (Fig. 28). In this form the seven discs can be freely cycled round the outer circle (which we hardly count as a move) or else a single disc may be slid across the bridge (remember that in the actual form of the puzzle the bridge is too short for several discs to traverse it at once). It doesn't really matter whether the disc we slide across the bridge goes upwards or downwards, since this has the same effect on the cyclic order, so we'll always slide our discs *downward*.

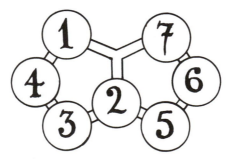

Figure 29. Swapping Two and Four.

If we think of the puzzle in this way and, starting from the home position, slide discs 2, 4, 2, 4, 2 down the bridge, we reach the position of Fig. 29 in which discs 2 and 4 have been interchanged and all the others are in their original places. Obviously we can interchange any pair of discs which are two places apart round the circle in this way. It's not hard to see how *any* desired rearrangement can be reached by a succession of such interchanges. For instance if we wanted to get

disc	1	2	3	4	5	6	7
to place	7	6	5	4	3	2	1

we might perform the interchanges of the following scheme

$$
\begin{array}{ccccccc}
1 & 2 & 3 & 4 & 5 & 6 & 7 \\
3 & 2 & 1 & 4 & 5 & 6 & 7 \\
3 & 2 & 5 & 4 & 1 & 6 & 7 \\
3 & 2 & 5 & 4 & 7 & 6 & 1 \\
3 & 4 & 5 & 2 & 7 & 6 & 1 \\
3 & 4 & 5 & 6 & 7 & 2 & 1 \\
5 & 4 & 3 & 6 & 7 & 2 & 1 \\
5 & 4 & 7 & 6 & 3 & 2 & 1 \\
5 & 6 & 7 & 4 & 3 & 2 & 1 \\
7 & 6 & 5 & 4 & 3 & 2 & 1
\end{array}
$$

Get 1 in position first,
then 2,
then 3,
then 4,
then 5 (6 and 7).

leading to a solution in which 45 discs have crossed the bridge. This method is not very efficient but it has the great advantage of providing an almost mechanical technique by which you can obtain any position. Can you find a shorter solution to the above problem?

All Other Courses for Point-to-Point

The history of the Fifteen Puzzle has been given too many times to bear further repetition here. Exactly half of the

$$15! = 1 \times 2 \times 3 \times \ldots \times 15 = 1\,307\,674\,368\,000$$

permutations (the so called *even* permutations) can be obtained. In technical language, the available permutations form the **alternating group**, A_{15}, whereas for the Lucky Seven Puzzle we have the full **symmetric group**, S_7, of $7! = 7 \times 6 \times 5 \times 4 \times 3 \times 2 \times 1 = 5040$ permutations.

You can make a puzzle of this type by putting counters on all but one of the nodes of any connected graph and then sliding them, point to point, always along an edge into the currently empty node. We can afford to ignore the *degenerate* cases, when your graph is a cycle, or is made by putting two smaller graphs together at a single node, because then the puzzle is trivial, or degenerates into the two smaller puzzles corresponding to the two smaller graphs.

Rick Wilson has proved the remarkable theorem that for every non-degenerate case but one we get either the full symmetric group (if some circuit is odd) or the alternating group (otherwise). The single exception is the graph of the **Tricky Six Puzzle** (Fig. 30) for which the group consists of all possible Möbius transformations

$$x \to \frac{ax+b}{cx+d} \pmod 5$$

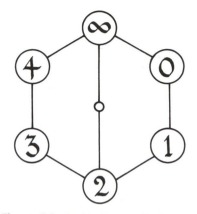

Figure 30. Rick's Tricky Six Puzzle.

Rubik's Hungarian Cube — Bűvös Kocka

The Hungarian words actually mean "magic cube". If you're crazy enough to get one of these you'll see that when it comes to you from the manufacturer it has just one color on each face (Fig. 31(a)) but your Hungarian cube is unlikely to stay in this beautiful state because you can rotate the nine little **cubelets** that make up any face (Fig. 31(b)) and so disturb the color scheme. For example, if you complete the turn started in Fig. 31(b), and then turn the top face clockwise you'll arrive at Fig. 31(c). After three more turns the colors are all over the place (Fig. 31(d)) and you'll find it very hard to recover the original arrangement; in other words to get each of the cubelets back into its own **cubicle**, *and* the right way round.

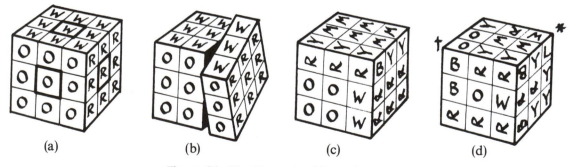

Figure 31. The Hungarian Magic Cube.

There are really two problems about this elegant little puzzle. The first is how its brilliant designer, Ernő Rubik, can possibly have managed to make all those motions feasible without all the cubelets falling apart. We'll leave that one to you! The other is, of course, to provide a method by which we can guarantee to get home from any position our friends have muddled the cube into.

Just How Chaotic Can the Cube Get?

At least there are six permanent landmarks: the cubelets at the centres of the faces always stay in their own cubicles although they may be rotated. We call these the **face cubelets** and have framed them in Fig. 31 (a). No matter how confused your cube looks, you can tell what the final color of each face should be, just by looking at the face cubelet at its centre. So, for instance, in Fig. 31 (d) we call the top face **white** even though only one third of it really is.

So you can work out the home cubicle for any cubelet by just looking at its colors and thinking which faces these belong to. For instance the LWO cubelet ∗ in Fig. 31(d) should end up at † (in our cube the colors opposite R, W, O are L, B, Y). We recommend the nervous novice always to hold the cube with its white face uppermost and then to take a careful note of the color of the bottom face, which we call the **ground** color.

Since the other 20 visible cubes are of two types,

> 8   **corner cubes**,   which have 3 possible orientations in their cubicles,
> and 12   **edge cubes**,   which have 2,

there are at most

$$3^8 \times 2^{12} \times 8! \times 12! = 519\,024\,039\,293\,878\,272\,000$$

conceivable arrangements. However, Anne Scott proved that only one-twelfth of this number, namely

$$43\,252\,003\,274\,489\,856\,000$$

are attainable.

Chief Colors and Chief Faces

These notions help us keep track of the orientations of cubelets, even when they're not in their home cubicles. We'll call the **chief face** of a *cubicle* the one in the top or bottom surface of the cube, if there is one, and otherwise the one in the right or left wall. The **chief color** of a *cubelet* is the color that should be in the chief place when the cubelet gets home. In other words White or the Ground color if possible, and otherwise the color that should end up in the left or right wall of the cube.

If a cubelet, no matter where it is, has its chief color in the chief face of its current cubicle we'll call it **sane** and otherwise **flipped** (if it's an edge cubelet) or **twisted** (if it's a corner one). There's only one way to make an edge-flip (e), but a corner may be twisted anticlockwise (a) or clockwise (c).

Now, as shown in Fig. 32, turning the top (or bottom) preserves the chiefness of every cubelet. Turning the front (or back) changes the chiefness at four corners and turning the left (or right) changes it at four corners and four edges. Since each turn flips an even number of edges, you can see that for attainable positions

> the total number of edge-flips
> will always be even.

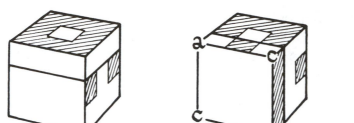
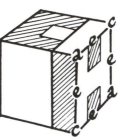

Figure 32. Changes in Chiefness.

And since each turn produces equal numbers of clockwise and anticlockwise twists

> the total corner twisting
> will always be zero, mod 3.

In computing corner twists we count $+1$ for clockwise and -1 for anticlockwise — of course three clockwise twists of a cubelet produce no effect. Finally, for reasons as in the Fifteen Puzzle

> the total permutation of all the
> 20 movable cubelets must be *even*.

An **even permutation** is one we might imagine making by an even number of interchanges.

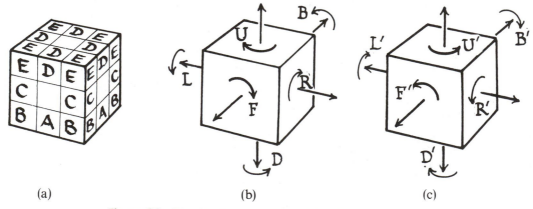

(a)   (b)   (c)

Figure 33. The Six Stages and Our Notation for the Moves.

Curing the Cube

Benson, Conway and Seal have simplified Anne Scott's proof that you really can get home from any position for which

  (i) the total edge flipping is zero, mod 2,

 (ii) the total corner twisting is zero, mod 3, and

(iii) the total permutation of all 20 movable cubes is even.

We have adapted our names for the moves (Fig. 33) so as to agree with David Singmaster's in the hope that a single notation will rapidly become universal. Note that the unprimed letters L,R,F,B,U,D, refer to *clockwise* turns, and the primed letters L′,R′,F′,B′,U′,D′ to *anticlockwise* ones. Our notation for the **slice moves** is illustrated in Fig. 34. Note that in these moves only the *middle* layer of the cube is turned. We shall also use the common notation in which, for example, X^2 means "do X twice" and X^{-1} means "undo X".

Figure 34. Slice Moves.

Our method has six stages which correspond roughly to the letters in Fig. 33(a).

A: Aloft, Around (Adjust) and About.   D: Domiciling the Top Edge Cubelets.
B: Bottom Layer Corner Cubelets.   E: Exchanging Pairs of Top Corners.
C: Central Layer Edge Cubelets.   F: Finishing Flips and Fiddles.

We've collected the figures for these stages in Fig. 35 for easy reference, so keep a finger on page 871.

Warning: Be very careful when applying this algorithm. Think of "tightening" or "loosening" a screw-cap, so that you never mistake a clockwise turn for an anticlockwise one, even from behind. Be aware at all times which way up you are holding the cube, and don't stop to think in the middle of a sequence of moves. Remember that if you make a tiny mistake you'll probably have to go all the way back to Stage A.

THE CUBE SELDOM FORGIVES!

A: Aloft, Around (Adjust) and About

Our first stage (Fig. 35A) gets the bottom edge cubelets (A in Fig. 33(a)) into their correct cubicles, the right way round. You bring the ground (= chief) color of such a cubelet into the topmost surface (Aloft) then turn the top layer Around to put this cubelet into the correct side wall which can be turned About to home the cubelet. Sometimes this disturbs a bottom edge cubelet that's already home, but this can be Adjusted by turning the appropriate side wall just before the About step.

B: Bottom Layer Corner Cubelets

Now, without disturbing the bottom layer edge cubelets, you must get the bottom layer corner cubelets home.

    If the cubelet that's to stand on the shaded square of Fig. 35B is in the top layer, turn the top layer until this cubelet's ground color is in one of the three numbered positions. Then do the appropriate one of

$$\text{B1}: F'U'F \quad \text{B2}: RUR' \quad \text{B3}: F'UF.RU^2R'$$

If the cubelet is already *in* the bottom layer, but wrongly placed, use one of these to put any corner cubelet from the top layer into its current position, thereby evicting it into the top layer. Then work as above to put it into the proper place. Repeat this procedure for the other three bottom layer corner cubelets.

C: Central Layer Edge Cubelets

This stage corrects the central layer edge cubelets without affecting the bottom layer.

    If the cubelet destined for the shaded cubicle of Fig. 35C is in the top layer, turn the top layer until you want to move this cubelet in one of the two ways of Fig. 35C (its side face will then be just above the face cubelet of the same color). Then do the appropriate one of

$$\text{C1}: URU'R'.U'F'UF \quad \text{C2}: U'F'UF.URU'R'$$

If the cubelet is already *in* the central layer, but wrongly placed, use one of these to evict it into the top layer. Then work as above. Repeat the procedure for the other three central layer edge cubelets.

D: Domiciling the Top Edge Cubelets

i.e. putting the top layer edge cubelets into their own home cubicles without as yet worrying about their orientations.

    You can do this by a sequence of swaps of adjacent edge cubelets as in Fig. 35D for which the moves are

$$UF.RUR'U'.F'$$

Of course you can first turn the top layer to reduce the number of swaps needed.

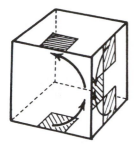 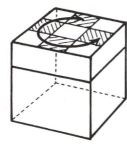 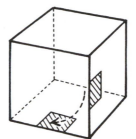 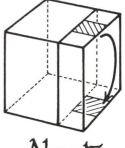

Aloft   Around   (Adjust)?   About

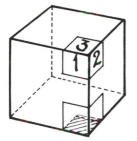 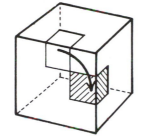 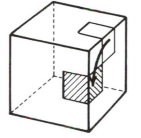 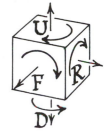

B1: F'UF
B2: RUR'
B3: F'UF.RU²R'

C1: URU'R'.UF'UF   C2: UF'UF.URU'R'

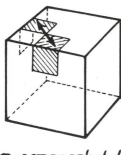 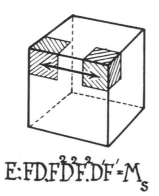 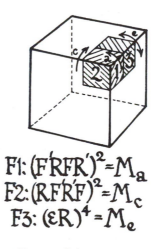 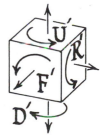

D: UF.RURU'.F'   E: FD.F'D²F.DF'=M_s

$F1: (F'RFR')^2 = M_a$
$F2: (RF'R'F)^2 = M_c$
$F3: (\mathcal{E}R)^4 = M_e$

Figure 35. Six Simple Stages Cure Chaotic Cubes.

E: Exchanging Pairs of Top Corners

Now you must get the top layer corner cubelets into their own cubicles by moves that, when they are finally completed, won't have affected the bottom two layers or moved the top layer edge cubelets. Usually you can do this in just two swaps of adjacent corners, but sometimes four will be needed.

Correct performance requires some care. Work out a pair of successive swaps of adjacent corner cubelets that will improve things. Then turn the cube until the first required swap is as in Fig. 35E and do our

$$\text{monoswap, } M_s = FD.F^2D^2F^2.D'F'$$

Then turn THE TOP LAYER *ONLY* to bring the second desired swap into the position of Fig. 35E, do another monoswap, and then return the top layer to its original position.

Since the bottom two layers are disordered by a single monoswap, but restored by a second one, it's important not to move these layers (by turning the cube, say) between the two mono-swaps of each pair.

F: Finishing Flips and Fiddles

Since every cubelet should now be in its own cubicle, the only remaining problems can be solved by edge-flips and corner twists in the top layer. To tackle any particular top layer cubelet, turn THE TOP LAYER *ONLY* to bring that cubelet into one of the two shaded cubicles of Fig. 35F and then, according as its white face is in position

1,	2	or	3

do our

| **anticlockwise monotwist,** | **clockwise monotwist** | or | **edge monoflip** |
| $M_a = (F'RFR')^2$ | $M_c = (RF'R'F)^2$ | | $M_e = (\varepsilon R)^4$ |

where ε is a slice move (Fig. 34).

Once again it's important not to move the bottom two layers by turning the cube between operations, since individual monotwists and monoflips affect these layers. However, the entire set of operations needed to correct the top layer will automatically correct the bottom two layers as well.

Explanations

Stage E works because our monoswap operation M_S leaves the top layer unchanged except for the desired swap of the two near corner cubelets, while two copies of the monoswap cancel ($M_S^2 = 1$).

So a sequence such as

monoswap, turn top clockwise, monoswap, turn top back

doesn't really disturb the bottom two layers, which "feel" only the two cancelling monoswaps. The top layer, however, effectively undergoes a swap of the two near corners followed by a swap of two right corners, which are brought into position by the first top turn and returned by the second.

Stage F works similarly because M_a, M_c and M_c have exactly the desired effects on the top layer, and enjoy the properties

$$M_e^2 = M_C^3 = 1, \quad M_c M_e = M_e M_c, \quad M_a = M_C^{-1}.$$

So Anne Scott's laws ensure that the bottom two layers feel a cancelling combination of operations, while the top layer undergoes the desired flips and twists.

Improvements

Our method is easy to explain, perform and remember, but usually takes more moves than an expert would. If you're prepared to take more trouble and have a rather larger memory, you can often shorten it considerably. For instance, the original monoflips and monotwists (due to David Seal and David Goto) are shorter:

$$m_e = R\varepsilon R^2 \varepsilon^2 R \qquad m_e^{-1} = R'\omega^2 R^2 \omega R' \qquad m_c = R'DRFDF' \qquad m_a = m_c^{-1} = FD'F'R'D'R$$

but with these you must always be careful to follow a mono-operation by the corresponding inverse one.

Explore the effects of the following moves, which many people have found useful. The first few only affect the top layer. Here and elsewhere we've credited moves to those who first told them to us. We expect that many facts about the cube were found by clever Hungarians long before we learnt of them. For the Greek letter slice moves, see Fig. 34.

David Benson's "special" $RUR2$. $FRF^2.UFU^2$
David Singmaster's "Sigma" $FURU'R'F'$
Margaret Bumby's top edge-tricycle $\beta U^{\pm 1}\alpha$. $U^2.\beta U^{\pm 1}\alpha$
Two more top edge-tricycles $U^2F.\alpha U\beta.U^2.\alpha U\beta.FU^2$; $FUF'UFU^2F'U^2$
Top corner tricycle $RU'L'UR'U'LU$
Clive Bach's cross-swap $(\alpha^2 U^2 \alpha^2 U)^2$
Kati Fried's edge-tricycle $\beta F^2 \alpha F^2$
Tamas Varga's corner tricycle $((FR'F'R)^3 U^2)^2$
Two double edge-swaps $(R^2 U^2)^3$; $((\alpha^2 U^2)^2$
Andrew Taylor's Stage C moves $F^2(RF)^2(R'F')^3$; $(FR)^3(F'R')^2F^2$
Other Stage C moves $FUFUF$. $U'F'U'F'U'$; $R'U'R'U'R'$. $URURU$

In the Extras you'll find lists of the shortest known words (improvements welcome!) to achieve any rearrangement, or any reorientation of the top layer. These are quoted from an algorithm due to Benson, Conway and Seal which guarantees to cure the cube in at most 85 moves (a half turn still counts as one move, but a slice counts as two). Morwen Thistlethwaite has recently constructed an impressive algorithm which never takes more than 52 moves.

Because there are 18 choices for the first move, but only 15 (non-cancelling) choices for subsequent ones, the number of positions after 16 moves is at most

$$18 \times 15^{15} = 7\,882\,090\,026\,855\,468\,750 < 43\,252\,003\,274\,489\,856\,000$$

proving that there are many positions that need 17 or more moves to cure. We can improve this to 18 moves by using the estimates $u_1 = 18$, $u_2 = 27 + 12u_1 = 243$, $u_{n+2} \le 18u_n + 12u_{n+1}$, which take into account relations like $LR = RL$.

Elena's Elements

Elena Conway likes making her cube into pretty patterns. Here are some ways she does this:

"4 Windows"	"6 Windows"	"Chequers"	"Harlequin"
$\alpha\gamma^2\beta\delta^2$	$\alpha\gamma\beta\delta$	$\alpha^2\gamma^2\varepsilon^2$	$\alpha\gamma\beta\delta\alpha^2\gamma^2\varepsilon^2$

"Stripey"	"Zigzag"	"4 Crosses"	"6 Crosses"
$(L^2F^2R^2)^2.LR'$	$(LRFB)^3$	$(LRFB)^3(FBLR)^3$ or $(\gamma^2L'\gamma^2R)^3$	$(\gamma^2L'\gamma^2R)^3(\alpha^2B'\alpha^2F)^3$

And try following "6 Crosses" with any of the earlier ones.

Are You Partial to Partial Puzzles?

It's interesting to see what you can do using only *some* of the available moves. You might restrict yourself to just a specified selection effaces, to half-turns, to slice moves, or to the **helislice moves** like LR. Mathematically these correspond to subgroups we call the 2-, 3-, 4- and 5-**face groups**, the **square group**, the **slice group** and the **helislice group**.

Beginners are recommended to stay in the slice group because they cannot get lost. From any position you can cure the edge-cubes in 3 slices, getting to "4 Windows" or "6 Windows" and so home in 4 more slices. Frank O'Hara has shown that in fact at most 5 slices are needed in all. The slice group has order $4^3.4!/2 = 768$ and the helislice group has order $2^{11}.3 = 6144$.

The 2-face group has been intensively studied by Morwen Thistlethwaite. It's interesting to notice that it involves both the lucky Seven Puzzle (on the edge cubelets that move) and Rick Wilson's Tricky Six Puzzle (on the corners).

Roger Penrose first proved that everything can be done using just 5 faces. David Benson has a simple proof:

$$RL'F^2B^2RL'.U.RL'F^2B^2RL' = D.$$

Other "Hungarian" Objects

A $2 \times 2 \times 2$ cube and $2 \times 3 \times 3$ "domino" have also been manufactured. Their design seems even more mysterious, although as puzzles they're much easier. One can *imagine* Hungarian tetrahedra, octahedra, dodecahedra, icosahedra, etc. Although, as far as we know, these have not all been manufactured nor completely solved, Andrew Taylor has found a neat proof that (for any choice of chief faces and colors)

the total permutation on edges and corners is even,
the number of edge-flips is even, and
the total corner twisting is zero, modulo the corner valence.

Despondent Domino dabblers should need but three little words (with effects):

$$X = EhEhEh \qquad Y = EcEhNcE \qquad Z = cYcYc$$
$$(28) \qquad (13)(26)(1'3')(2'6') \qquad (13)(26)$$

(c is a clockwise $\frac{1}{4}$-turn of the top; h,E,N $\frac{1}{2}$-turns of top, East, North).

A Trio of Sliding Block Puzzles

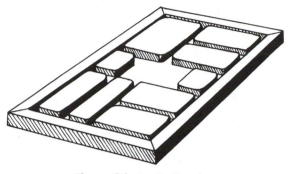

Figure 36. Dad's Puzzler.

Dad's Puzzler (Fig. 36) is unfortunately almost the only sliding block puzzle that's generally available from toy stores, although it goes under many different names. The problem is to slide the pieces without lifting any out of the tray, until the 2 × 2 square arrives in the lower left hand corner. Fifty years ago the puzzle represented Dad's furniture-removing difficulties, and the 2 × 2 block was the piano; at other times it has been depicted as a pennant, a car, a mountain, or space capsule but the puzzle has remained unchanged, probably for a hundred years. Some more enterprising manufacturer should sell a set containing one 2 × 2, four 1 × 1 and six 2 × 1 pieces which can be used either for Dad's Puzzler or for the following more interesting puzzles.

In the **Donkey** puzzle the initial arrangement is as in Fig. 37(a) and the problem is to move the 2 × 2 square to the middle of the bottom row. The name arises from the picture of a red donkey which adorned the 2 × 2 square in the original French version (L'Âne Rouge, which probably goes back to the last century) but we think that our choice of starting position already looks quite like a donkey's face.

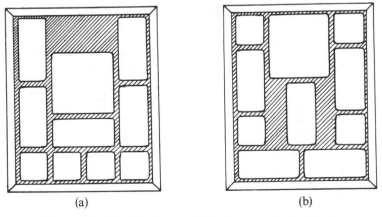

(a)   (b)

Figure 37. The Donkey and The Century (and a Half).

The **Century Puzzle**, published for the first time in *Winning Ways*, was discovered by one of us several years ago as a result of a systematic search for the hardest puzzle of this size. Start from Fig. 37(b) and, as in the Donkey, end with the 2 × 2 block in the middle of the bottom row. Or, if you're a real expert, you might try the **Century-and-a-Half Puzzle** in which you're to end in the position got by turning Fig. 37(b) upside-down.

Tactics for Solving Such Puzzles

As in our previous sliding puzzles the basic idea is to see what can be done while quite a lot of the pieces are kept fixed. In all three of these examples one occasionally sees one of the configurations of Fig. 38 somewhere, and any of these can be exchanged for any other, moving

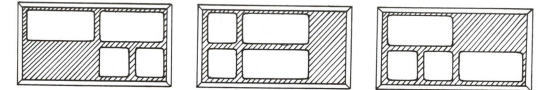

Figure 38. A Micropuzzle.

only the pieces in the area shown. They form a kind of micro-puzzle within the larger one. Figure 40 is a complete "map" of Dad's Puzzler showing how it consists of a dozen of these micro-puzzles joined by various paths of moves that are more or less forced. Using this map, you'll find it easy to get from anywhere to anywhere else.

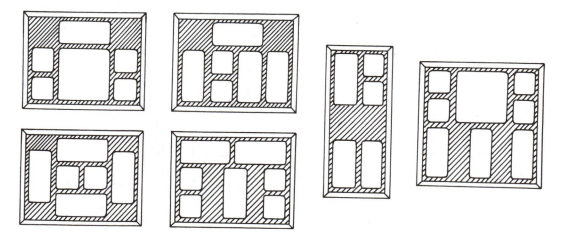

Figure 39. Micro- and Mini-puzzles Found in Donkey and Century.

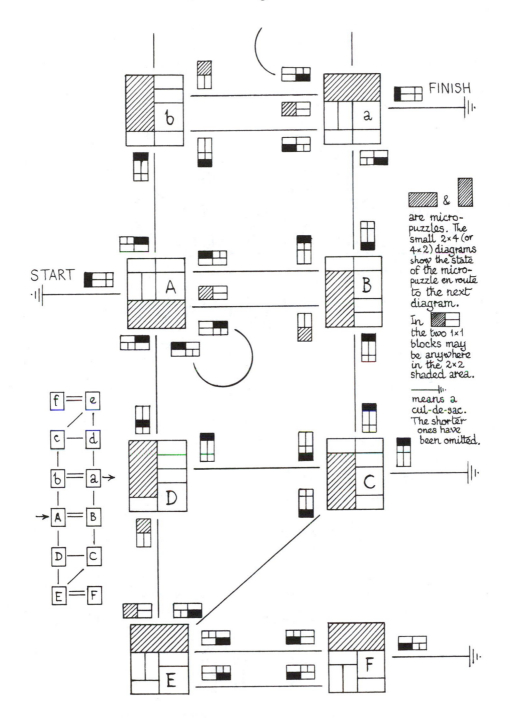

Figure 40. Map of Dad's Puzzler.

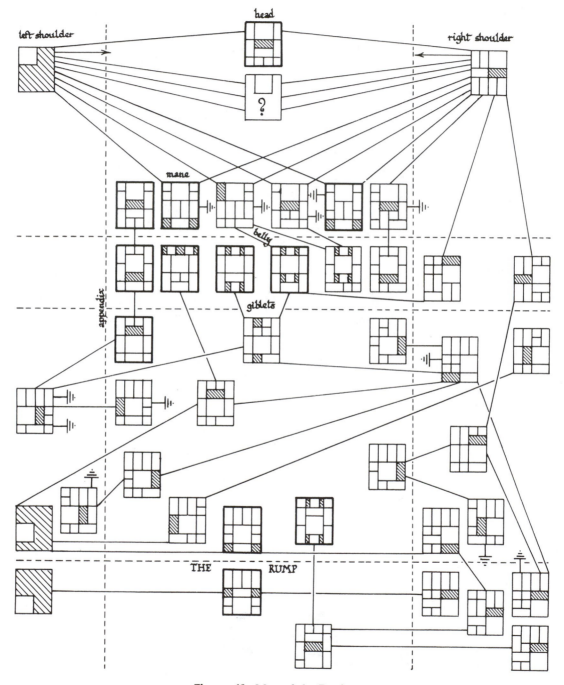

Figure 41. Map of the Donkey.

In the Donkey and Century puzzles there are several micro- and mini-puzzles: see what moves you can make inside the regions shown in Fig. 40. The Century and Donkey puzzles will never become easy but it will help if you become an adept at these minipuzzles. Figure 41 is our map of the Donkey. The positions are classified according to the location of the 2×2 square and in most cases we have only drawn one of a left-right mirror-image pair. Some unimportant culs-de-sac will be found in the directions indicated by the signs ⊣‖⊪, and the rectangle containing (?) represents many positions connected to the left and right shoulders. The arrows indicate other connexions to the shoulders. Left-right symmetric positions are boldly bordered.

The Century puzzle is very much larger, and we need more abbreviations to draw its map within a reasonable compass. The positions are best classified by the position of the large square together with information about which of the two horizontal pieces should be counted as "above" or "below" the square. We remark that in Fig. 42 both horizontal pieces should be counted as *below* the square despite their appearance, because the only way to move these pieces takes the horizontals *down* and the square *up*.

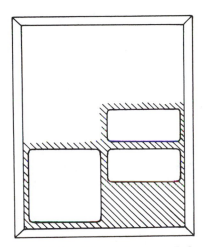

Figure 42. The Two Horizontal Pieces are Below the Square!

The key to the puzzle is to find one of the two possible **narrow bridges** in the map at which the first horizontal piece changes from *below* the square to *above* it. In fact it's best to think out the possible configurations in which this can happen and then work the puzzle backwards and forwards from one of these. Very few people have ever solved the puzzle by starting at the initial configuration and moving steadily towards its end. A much abbreviated map appears as Fig. 43.

Our maps were prepared with much help from some computer calculations made by David Fremlin at the University of Essex, who found incidentally that the Donkey pieces may be placed in the tray in 65880 positions and the Century pieces in 109260 ways. Although the Century puzzle can be inverted (this is our Century-and-a-Half problem) Fremlin's computer found that the Donkey cannot. It would be nice to have a more perspicuous proof of this.

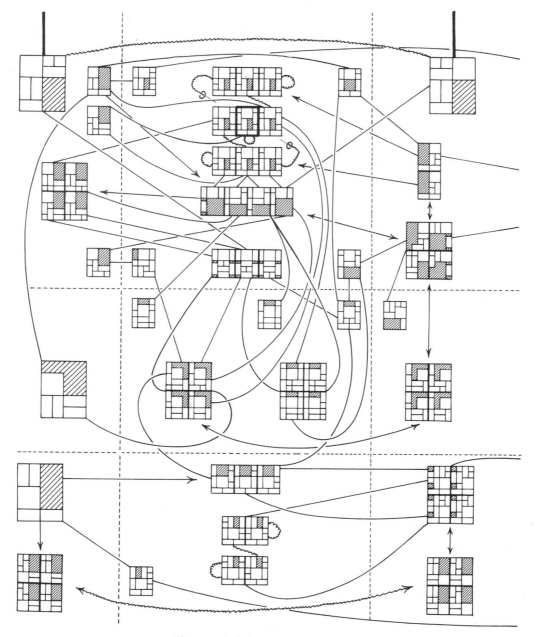

Figure 43. Map . . . .

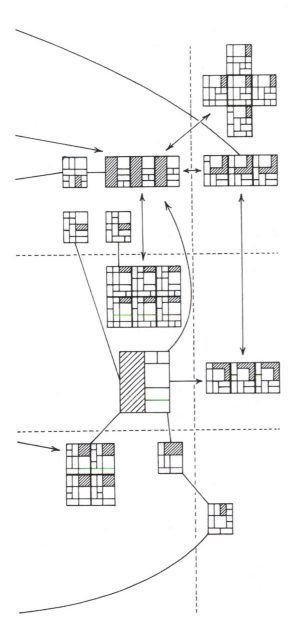

The starting position is heavily outlined; see centre column of opposite page, near top

Positions are classified thus:

    overleaf: Square "between" horizontals.
    opposite: Two verticals left, one right.
    this page: Three verticals left.

They are further classified according to the location of the square:

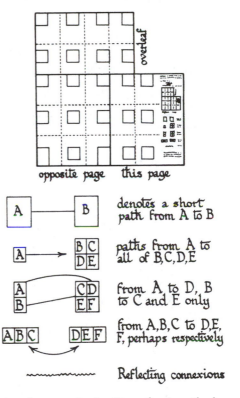

opposite page   this page

A ——— B   denotes a short path from A to B

A ——→ [BC DE]   paths from A to all of B,C,D,E

[A B] ⌒ [CD EF]   from A to D, B to C and E only

[ABC] ⌒ [DEF]   from A,B,C to D,E, F, perhaps respectively

〜〜〜〜〜   Reflecting connexions

The "narrow bridges" are the two thick connexions between the top of the opposite page and the left of overleaf.

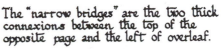

.... of the ....

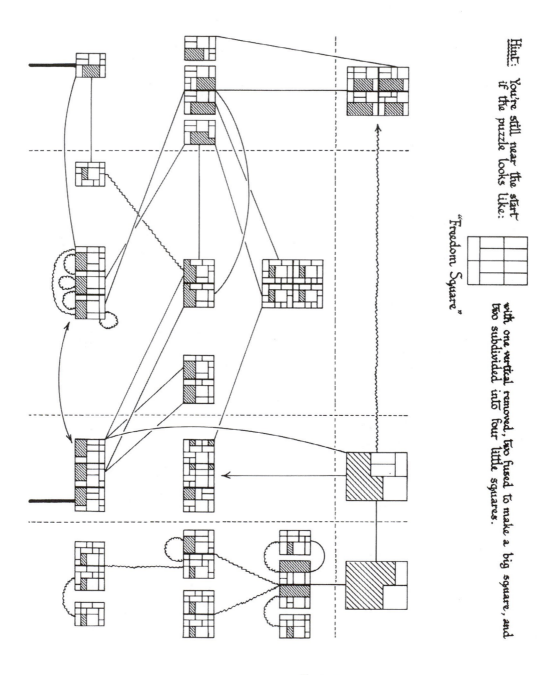

Hint: You're still near the start if the puzzle looks like:

"Freedom Square"

with one vertical removed, two fused to make a big square, and two subdivided into four little squares.

.... Century.

Counting Your Moves

It's customary to follow Martin Gardner and declare that any kind of motion involving just one piece counts as a single move. It takes 58 moves to solve Dad's Puzzler and 83 to solve the Donkey. How many do you need to solve the Century puzzle? And how many for the Century-and-a-Half?

Paradoxical Pennies

You tell me your favorite sequence of three Heads or Tails and then I'll tell you mine. We then spin a penny until the first time either of our sequences appears as the result of three consecutive throws. I bet you 2 to 1 it's mine!

The graph

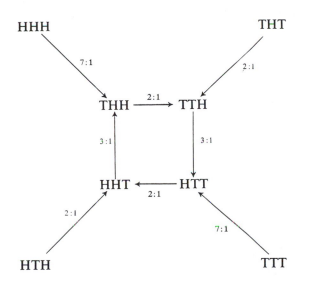

shows the sequence I'll choose for each possible sequence of yours, together with the odds that I win. You'll see that it's always at least 2 to 1 in my favor.

Here's a rule for computing the odds. Given two Head-Tail sequences a and b of the same length, n, we compute the **leading number**, aLb, by scoring 2^{k-1} for every positive k for which the last k letters of a coincide with the first k of b. Then we can show that the odds, that b beats a in Paradoxical Pennies, are exactly

$$aLa - aLb \quad \text{to} \quad bLb - bLa.$$

Leo Guibas and Andy Odlyzko have proved that, given a, the best choice for b is one of the two sequences obtained by dropping the last digit of a and prefixing a new first digit. Notice the paradoxical fact that in the length 3 game:

THH beats HHT beats HTT beats TTH beats THH.

Paradoxical Dice

You can make three dice , A, B, C, with a similar paradoxical property, using the magic square:

	D	E	F
A	6	1	8
B	7	5	3
C	2	9	4

Each die has the numbers of one row of the square on its faces (opposite faces bearing the same number). For these dice

$$\text{A beats B beats C beats A,}$$

all by 5 to 4 odds! Similarly for the three dice, D, E, F, obtained from the columns. The only other paradoxical triples of dice using the same numbers are those obtained from A, B, C by interchanging 3 with 4 and/or 6 with 7. These interchanges improve the odds.

It's possible to put positive integers on the faces of two dice in a unique non-standard way that gives the same probability for each total as the standard one. Algebraically, the problem reduces to factorizing

$$x^2 + 2x^3 + 3x^4 + 4x^5 + 5x^6 + 6x^7 + 5x^8 + 4x^9 + 3x^{10} + 2x^{11} + x^{12}$$

into the form $f(x)g(x)$ with $f(0) = g(0) = 0$ and $f(1) = g(1) = 6$. The two factorizations are

$$(x + x^2 + x^3 + x^4 + x^5 + x^6)^2 \quad \text{and}$$
$$(x + 2x^2 + 2x^3 + x^4)(x + x^3 + x^4 + x^5 + x^6 + x^8),$$

so the new pair of dice have the numbers

$$1, 2, 2, 3, 3, 4 \quad \text{and} \quad 1, 3, 4, 5, 6, 8.$$

More on Magic Squares

It's an old puzzle to arrange the numbers from 1 to n^2 in an array so that all the rows and columns and both the diagonals have the same sum, which turns out to be $1/2n(n^2 + 1)$. The only 3×3 magic square (see the last section), often called the Lo-Shu , was discovered several dynasties ago by the Chinese. We also used it in Chapter 22. In 1693 Frenicle de Bessy had worked out the 880 magic squares of order 4. In this section we'll show you how to find all these.

It's handy to subtract 1 from all the numbers, because the numbers 0 to 15 are closed under nim-addition. With this convention the magic sum is 30. We shall call a square **perfect** if we can nim-add *any* number from 0 to 15 to its entries and still obtain a magic square; if only 1/2 of these additions are possible we'll call it 1/2-**perfect**, and so on. Since nim-adding 15 is the same as complementing in 15, it always preserves the magic property, showing that *every* square is at least 1/8-perfect. We shall also classify the squares by the disposition of complementary pairs as in Fig. 44.

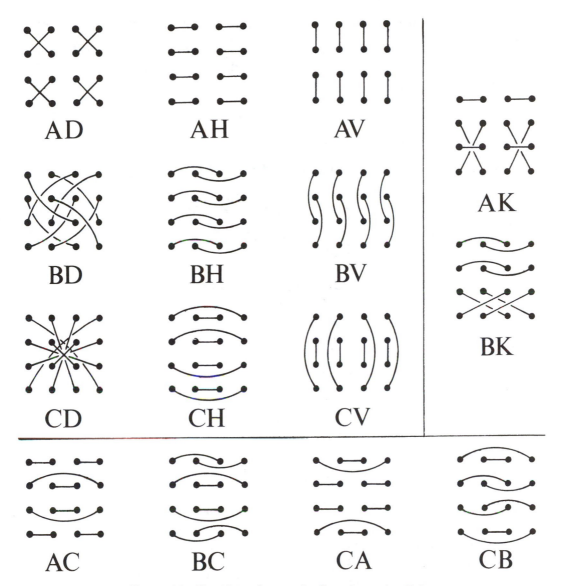

Figure 44. Classifying Squares by Complementing Pairs.

There are essentially just three ways to write the numbers from 0 to 15 as an addition table:

0	1	2	3	0	1	4	5	0	2	4	6
4	5	6	7	2	3	6	7	1	3	5	7
8	9	10	11	8	9	12	13	8	10	12	14
12	13	14	15	10	11	14	15	9	11	13	15

but you can then freely permute the rows and columns in any of these. Take any table obtained
in this way, say

15	11	14	10
13	9	12	8
7	3	6	2
5	1	4	0

Apply the interchanges indicated by our **Quaquaversal Quadrimagifier**

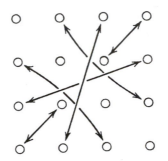

and you get the magic square on the left in:

15	2	1	12		16	3	2	13
4	9	10	7		5	10	11	8
8	5	6	11		9	6	7	12
3	14	13	0		4	15	14	1

Adding I to this particular example we obtain the right hand square which features in Albrecht
Dürer's famous self-portrait, *Melencolia I*, in which the boxed figures indicate the date of the
work. In this case complementary numbers appear according to the scheme called CD in
Fig. 44, and so this square is called Central Diagonal.

By applying the Quaquaversal Quadrimagifier to the other forms of addition table we can
get 432 essentially different perfect magic squares. The complementary pairs enable us to
classify these as:

48 Adjacent Diagonal (AD),

48 Broken Diagonal (BD),

48 Central Diagonal (CD),

96 Adjacent Horizontal (AH) or Adjacent Vertical (AV),

96 Broken Horizontal (BH) or Broken Vertical (BV), and

96 Central Horizontal (CH) or Central Vertical (CV).

Because we don't count squares as different when they are related merely by a reflexion or a rotation of the diagram, we must regard Adjacent-Horizontal and Adjacent-Vertical squares as the same type. You can find out what type your square will be by looking at the position occupied by the complement of the addition table's leading entry before Quadrimagification:

∗	○	○	○
○	AD	CH	BV
○	CV	BD	AH
○	BH	AV	CD

Now take the above 96 Central-Horizontal squares and apply the flip operation

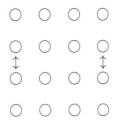

and you'll get 96 more Central-Horizontal squares. All squares so far found are perfect.

There are 112 more Central-Horizontal squares that are only 1/4 -or 1/8-perfect. They can be found by taking any of the seven squares:

```
 6  10   5   9         14   2  13   1                              10   1  14   5
13  12   3   2   a₈      5   4  11  10                c↗          13   8   7   2
 0   7   8  15          8  15   0   7      0  13   2  15            4  15   0  11
11   1  14   4          3   9   6  12     11   8   7   4            3   6   9  12
        ↓d                     ↓d         14   3  12   1                  ↓c
12   5  10   3         13   4  11   2      5   6   9  10           12  11   4   3
11   9   6   4   a₁     10   8   7   5                c↘           1   8   7  14
 0  14   1  15          1  15   0  14                              2   5  10  13
 7   2  13   8          6   3  12   9                              15   6   9   0
```

$\frac{1}{4}$-perfect $\frac{1}{8}$-perfect

and applying any combination of the four operations:

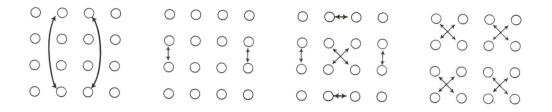

Now take the 14 squares of Fig. 45 and apply any combination of complementation and the last three of our operations and you'll get a total of 224 squares, 56 of each of the types

> Adjacent Central (AC),
> Broken Central (BC),
> Central Adjacent (CA), and
> Central Broken (CB).

```
 6  9  4 11      14  1 12  3       2 13  0 15      10  5  8  7
 8 14  1  7  a₈   0  6  9 15  b   12  6  9  3  a₈   4 14  1 11
 3  5 10 12      11 13  2  4      11  1 14  4       3  9  6 12      14  1  8  7
13  2 15  0       5 10  7  8       5 10  7  8      13  2 15  0       2 12  3 13
   ↓d               ↓d               ↓d               ↓d             5 11  4 10
12  3  8  7      13  2  9  6       4 11  0 15       5 10  1 14       9  6 15  0
 1 13  2 14  a₁   0 12  3 15       9 12  3  6  a₁   8 13  2  7          ↓d
 6 10  5  9       7 11  4  8       7  2 13  8       6  3 12  9      13  2  1 14
11  4 15  0      10  5 14  1      10  5 14  1      11  4 15  0       4  9  6 11
   ↓d               ↓d               ↓d               ↓d            10  7  8  5
 9  6  1 14      11  4  3 12       8  7  0 15      10  5  2 13       3 12 15  0
 2 11  4 13  a₂   0  9  6 15       3  9  6 12  a₂   1 11  4 14
12  5 10  3      14  7  8  1      14  4 11  1      12  6  9  3
 7  8 15  0       5 10 13  2       5 10 13  2       7  8 15  0
```

$\underbrace{\hspace{8cm}}_{\frac{1}{2}\text{-perfect}}$ $\underbrace{\hspace{2cm}}_{\frac{1}{4}\text{-perfect}}$

Figure 45. Adjacent and Broken Central and Central Adjacent and Broken Squares.

There remain only 16, rather irregular, squares to be found. You can get them by applying any combination of complementation and the last *two* of our operations to the two 1/2-perfect squares

$$
\begin{array}{cccc}
1 & 14 & 9 & 6 \\
10 & 3 & 4 & 13 \\
7 & 8 & 15 & 0 \\
12 & 5 & 2 & 11
\end{array}
\xrightarrow{d}
\begin{array}{cccc}
2 & 13 & 3 & 12 \\
5 & 6 & 8 & 11 \\
14 & 1 & 15 & 0 \\
9 & 10 & 4 & 7
\end{array}
$$

and they're 8 each of the types

Adjacent Knighted (AK),

Broken Knighted (BK).

There are various permutations of the 16 numbers that occasionally lead from one magic square to another, namely

a_n: nim-*add n*, for example $a_6 = (0\ 6)(1\ 7)(2\ 4)(3\ 5)(8\ 14)(9\ 15)(10\ 12)(11\ 13)$

b: the *big* swap $(0\ 12)(1\ 13)(14\ 2)(15\ 3)$

c: *circle* $(0\ 10\ 12)(1\ 11\ 13)(14\ 4\ 2)(15\ 5\ 3)$

d: *double*, mod 15 $(1\ 2\ 4\ 8)(3\ 6\ 12\ 9)(5\ 10)(7\ 14\ 13\ 11)$

and we've indicated some of these in the figures.

The Magic Tesseract

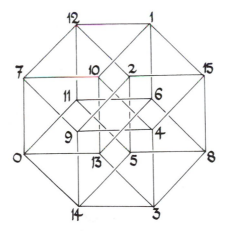

We'll leave it to you to rediscover the many remarkable relations between the 48 BD squares, sometimes called **pandiagonal** or **Nasik** squares, and our **Magic Tesseract** in which the vertices of every square add to 30. By projecting this along three different directions, you can find three magic cubes in which each face adds to 14. These are the duals of the three octahedral dice found by Andreas and Coxeter. Alternate vertices in the magic tesseract are the odious and evil numbers , and if you replace each odious number by its opposite (nim-sum with 15) you'll see how the tesseract was made.

Adams's Amazing Magic Hexagon

Starting from the pattern

```
                    1
              2           3
          4         5           6
              7           8
          9         10          11
              12          13
          14        15          16
              17          18
                    19
```

can you reorder the numbers from 1 to 19, taking less than 47 years, so that all five rows in each of the three directions have the same sum?

Strip-Jack-Naked, or Beggar-My-Neighbour **1

Another problem that took almost 47 years to solve concerns this old children's game. Each of the two players starts with about half of the cards (held face-down), which they alternately turn over onto a face-upwards "stack" on the table, until one of them (who's now "the commander") first deals one of the "commanding cards" (Jack, Queen, King, or Ace).

After one of these has been dealt, the other player (now "the responder") turns over cards continuously until EITHER **2 a new commanding card appears (when the players change roles **3) or respectively 1, 2, 3, or 4 non-commanding cards have been turned over. In the latter case, the commander turns over the stack and ajoins it to the bottom of his hand. The responder then starts the formation of a new stack by turning over his next card, and play continues as before.

A player who acquires all the cards is the winner and in real games, it seems that someone always does win. The interesting mathematical question, posed by one of us many years ago, was "is it really true that the game always ends?" Marc Paulhus has recently found the answer to be "no!". About 1 in 150,000 games (played with the usual 52 cards) goes on forever.

We are fairly confident that no one person has played the game anything like that number of times, so the chance (with random shuffling) of experiencing a non-terminating game in a lifetime's play must be very small indeed.

Just as surely, however, the total number of times this game has been played by the World's **4 children must be significantly larger than 150,000, so many of them will have been theoretically non-terminating ones. We imagine, though, that in practice most of them actually did terminate because someone made a mistake.

The Great Tantalizer

This is a tantalizing puzzle which surfaces every now and then with a new alias. We've chosen one of the older names. An early American version was the Katzenjammer puzzle, but most recently it has emerged under yet another name, Instant Insanity. The manufacturers seem to be very good at selecting new names, but they never change the underlying puzzle.

The problem

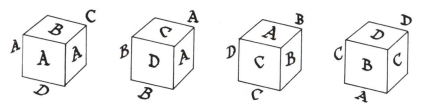

Figure 46. Pieces for The Great Tantalizer.

is to assemble the four cubes of Fig. 46 (in which the outer letters refer to the hidden faces) into a vertical 1 × 1 × 4 tower in which each wall displays all four "colors", A, B, C, D. If you don't go instantly insane on playing with the cubes, you'll probably be greatly tantalized by them.

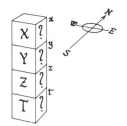

Figure 47. The Tantalizer Solved?

T.H. O'Beirne seems to have been the first to publish a general way of solving such problems and we think his solution is still the best. Let's imagine the problem solved and concentrate on the North and South walls of the tower (Fig. 47). Then X, Y, Z, T will be A, B, C, D in some order, as will x, y, z, t. Write the four letters A, B, C, D on a piece of paper and join

X to x, Y to y, Z to z and T to t.

What you'll get will probably be a way of joining ABCD into a circuit, but it might perhaps be several circuits which together include each letter just once. For example if

X Y Z T x y z t

are

A B C D D C A B

we get the single circuit

while if they were

$$A\ B\ C\ D\ D\ A\ C\ B$$

you'd get two circuits of different lengths

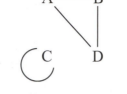

There will be a similar circuit, or system of circuits, for the East-West walls. Each of the two systems will contain every vertex just once and have one edge for each cube.

It's now easy to solve the puzzle by drawing the following graph (Fig. 48). The vertices of the graph are the colors A, B, C, D and the ith cube yields three edges labelled i joining pairs of vertices corresponding to its pairs of opposite faces. All you have to do is to select from this graph the two separate systems of circuits which each use all four numbers and all four vertices just once.

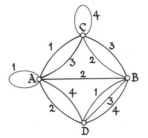

Figure 48. Solving the Tantalizer.

What are the possibilities for such circuit systems in the example? By considering each possibility

$$1111,\quad 211,\quad 22,\quad 31,\quad 4$$

for the circuit lengths, you'll rapidly conclude that both systems must consist of a single 4-circuit which can only use the letters in the cyclic order ACBD. There is only one way of selecting two such systems without using any edge twice:

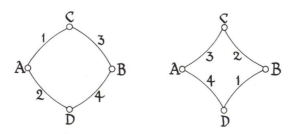

So the Great Tantalizer has a unique solution (up to reordering the cubes and rotating or inverting the whole tower). You can get it by pushing the cubes of Fig. 46 together left to right and tipping the result on end.

O'Beirne takes as his basic example a five cube puzzle of this type which dates from the first World War (Fig. 49) and uses the flags of the allies Belgium, France , Japan, Russia and the United Kingdom. You might like to check his assertion that this has just two essentially different solutions.

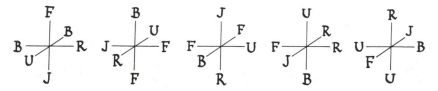

Figure 49. The "Flags of the Allies" Puzzle.

Polyominoes, Polyiamonds and Searching Policy

A domino is made of two squares stuck together, so S.W. Golomb has suggested the words tromino, tetromino, etc. for the figures that can be made by sticking 3, 4, or more equal squares together. He has registered the particular names pentomino (5 squares) and polyomino (*n* squares) as trade-marks. Unfortunately few of the puzzles that have been proposed have hidden secrets, so they yield to nothing better than trial and error (or systematic search). As Rouse Ball says about Tangrams in early editions of *Mathematical Recreations* and *Essays*, "the recreation is not mathematical and I reluctantly content myself with a bare mention of it".

Here is the type of puzzle that arises. Up to rotations and reflexions there are just 12 pentominoes, for which you'll find our naming system in Chapter 25, with a total area of 60 square units. Which of the candidate rectangles

$$3 \times 20 \quad 4 \times 15 \quad 5 \times 12 \quad 6 \times 10$$

can be packed with them? Figure 50 shows a way of solving two of these problems at once, and also, if the pieces are regarded as made of five cubes each, of packing the $2 \times 5 \times 6$ box (they

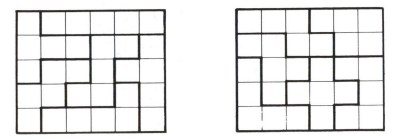

Figure 50. Packing Pentominoes.

will also pack a $3 \times 4 \times 5$ box). Such problems are peculiarly susceptible to idle computers and the 6×10 pentomino rectangle was one of the first to be tackled in this way when C.B. Haselgrove found its 2339 solutions in 1960.

Noting that two equilateral triangles can form a diamond, T.H. O'Beirne has proposed the terms triamond, etc., for figures made from three of more. Counting reflexions as distinct this time we find there are 19 hexiamonds, named in Fig. 51, which will pack into the shape of Fig. 52 in many thousands of different ways. The packing shown in the figure is the most symmetric about the North-South line. (See page 920.) We'd like to see a similarly symmetric one for the East-West line.

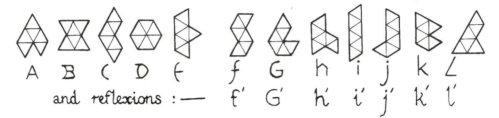

Figure 51. The Nineteen Hexiamonds.

This prompts a few remarks about sensible search procedures when solving puzzles or finding strategies for games that may be too large for complete discussion. Even when you have a large computer it's wise to have some idea where to look. Symmetry is usually a valuable consideration. For instance the (nearly) left-right symmetric solutions of the hexiamond

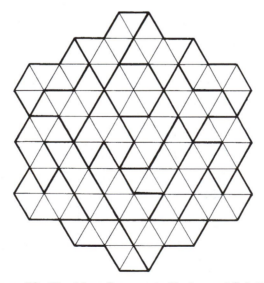

Figure 52. The Most Symmetric Hexiamond Solution?

puzzle admittedly form only a small corner of the space to be searched, but this one is likely to be a profitable one because the constraints on opposite sides of the board are satisfied simultaneously. However, symmetry is not the only consideration. In analyzing a game it's wise to try to find out what the players are really fighting for (the game's hidden secrets). For example the French Military Hunt game on the Small Board is small enough that you can give an exhaustive analysis without needing to understand what's really going on. But when you've discovered that the players are really just fighting over the opposition you can extend the analysis to much larger boards for which a complete analysis would be prohibitive, even by computer.

Many of the analyses in *Winning Ways* were found in this way. Only when we realized that Dots and Boxes was really more concerned with parity than with box counting were we able to make any headway. And it's impossibly complicated to evaluate a reasonably sized position in Hackenbush Hotchpotch exactly, but we got a head start when we realized that often the atomic weight was the only thing that really mattered. In Peg Solitaire the hidden secret turned out to be the notion of balance represented by α and β in Chapter 23.

Even though polyomino type problems may have no hidden secrets, some people are much better at them than others because they subconsciously search in more likely places. Experienced polyominists don't undo their good work by repeatedly starting from scratch but keep most of the puzzle in place while fiddling with just a few pieces at any time. When they've found one solution, they can usually transform it into others by similar manipulation. For example, from Fig. 50 you can obtain another solution by repacking pentominoes R and S, and in Fig. 52 we can interchange the two (f, h) pairs or rotate the central (A,D,E, j, j')hexagon.

Exercise for Experts: For what values of n can you pack n^2 copies of hexiamond A into a replica of A on n times the scale?

Alan Schoen's Cyclotome

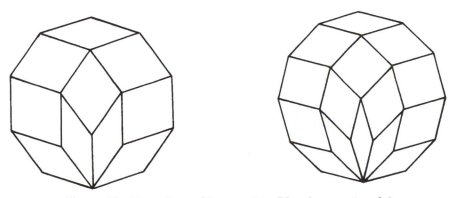

Figure 53. Dissections of $2n$-gons into Rhombs, $n = 5$ and 6.

Alan Schoen is patenting the interesting sequence of puzzles he derived from the well-known dissections of $2n$-gons into $\binom{n}{2}$ rhombs of angles $\pi k/n$, $1 \le k \le n - 1$ (Fig. 53). He takes one of each of the $\lfloor n/2 \rfloor$ shapes of rhomb and one of each of the shapes you can make by joining two rhombs in every possible way to form a hexagon. The hexagon must not contain a straight angle, since he observes that no packing of rhombs in the $2n$-gon contains a pair of parallel edges, except those which form the rungs of the "ladders" which run between each pair of opposite sides in every packing. This non-convexity condition is similar to that imposed by Piet Hein in designing the Soma pieces, but here it arises naturally. Reflexions, are not counted as different. This set of rhombs and hexagons (cyclotominoes?) will pack into the original $2n$-gon. In fact for

$$\begin{array}{rccccll} n = & 2 & 3 & 4 & 5 & \text{and} & 6 \\ \text{there are} & 1 & 1 & 3 & 14 & \text{and} & \text{more than } 150 \end{array}$$

essentially different packings. Schoen gave one of us a set of pieces for $n = 8$ and we were able to assemble them as in Fig. 54. We've numbered the pieces with the values of k, where $\pi k/n$ is the smaller angle of the rhomb. Where two shapes of piece are made from the same pair of rhombs, the one with the straighter reflex angle has its digits in natural order.

Solutions can be obtained from one another much as in O'Beirne's Hexiamond, or as on our Somap. In Fig. 54 the pieces 4 and 22 may be rotated or exchanged with 2 and 24, which in turn can be rotated or reflected. After this exchange, with 2 touching 11 and 34, we have a rotatable decagon, 1,2,34,11,32,3 & 4 of which the last four pieces form a rotatable octagon. As 3 & 4 are contiguous, they will exchange with 34, after which 4 & 1 and 2 & 3 are contiguous, and will swap with 14 and 23. After the original exchange, 2 may instead have two sides in common with 13 and these two will rotate, after which 21 and 12 may be interchanged if 1 & 2 are moved as well. Or again, 2 may touch 23 & 24, so that after the 34 exchange, 2 & 3 will swap with 32, and then 2 & 11 form a symmetric hexagon. And so on and on, yielding well over a hundred solutions.

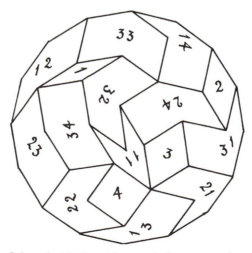

Figure 54. Schoen's 16-piece 16-gon. A Century or So of Solutions.

How many pieces are there in a set of cyclotominoes? According as $n = 2m$ or $2m + 1$, there are $m^2 - m$ or m^2 hexagons , and m rhombs in either case, so there are m^2 or $m^2 + m$ altogether. You can use sets for a variety of games and puzzles, ranging from Tangram-like pictures (Fig. 55) to quite sophisticated packing problems. It's early to say if these last contain any hidden secrets (though Alan Schoen has noted the one about parallel edges); there's perhaps a better chance since there is more structure in the shapes than there was in polyominoes and polyiamonds.

Many pleasing patterns can be produced: for example, take r^2 sets of pieces and pack them in nesting $2n$-gons of side lengths $1,2,\ldots,r$.

The exponential difficulty of this sequence of puzzles prompts us to add another remark about searching. A typical combinatorial puzzle or search of "size" n takes something like $n!$ trials to complete, and this is much more like n^n than c^n, no matter how big you take c. On the other hand the number of solutions may only be c^n, and while this goes up fast, your chance of finding one of them is only $(c/n)^n$ and this gets very small very fast as soon as n is bigger than c.

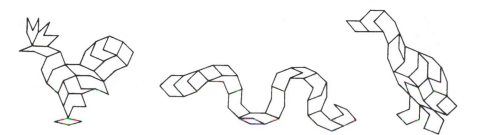

Figure 55. Schoen-Shapes Made with a Sixteen Set: Rooster, Serpent and Gosling.

MacMahon's Superdominoes

In his *New Mathematical Pastimes*, MacMahon proposed a different kind of generalized domino, got by dividing a regular polygon into colored triangles. We'll discuss just two examples. If we use just four colors, there are exactly 24 ways of coloring a triangular superdomino, and the standard problem is to pack these into a regular hexagon with an all black perimeter and adjacent colors alike, as in Fig. 56.

In this case it's hard to keep the secret hidden for very long. There are barely enough black edges to go round, and once you've found a suitable arrangement for them the rest is fairly easy.

When we consider the 24 three-colored square superdominoes, with which the usual problem is to make a 4×6 rectangle under similar conditions, the black edge problem is much more subtle. It can be shown that every solution to this problem has a column of four squares in which every horizontal edge is black (the **ladder**) . In Fig. 57(a) the ladder occupies the second column and in Fig. 57(b) it occupies the third. In the Extras you'll find every possible configuration for the black edges.

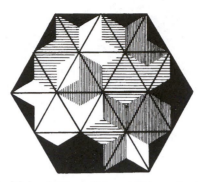

Figure 56. MacMahon's Four-Colored Triangular Superdominoes.

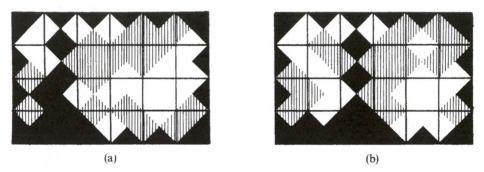

(a)   (b)

Figure 57. Three-Colored Square Superdomino Solutions Showing the Ladder.

MacMahon's superdomino problems can be made into jigsaw puzzles by using differently shaped edges in place of colors. Thus for the three colors in MacMahon's square problem one can use either the three edge shapes of Fig. 58(a) or those of Fig. 58(b) (which alter the matching condition).

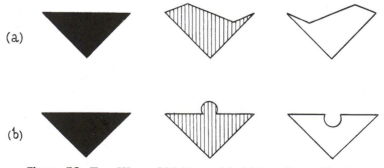

Figure 58. Two Ways of Making a MacMahon Jigsaw Puzzle.

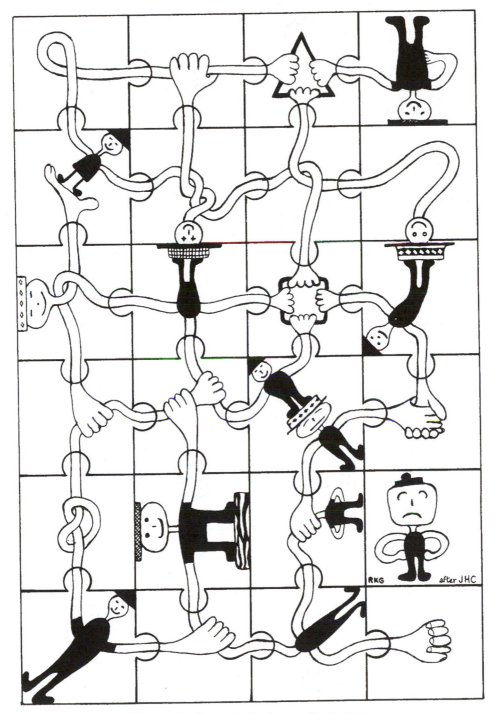

Figure 59. Conway's Christmas Card, 1968.

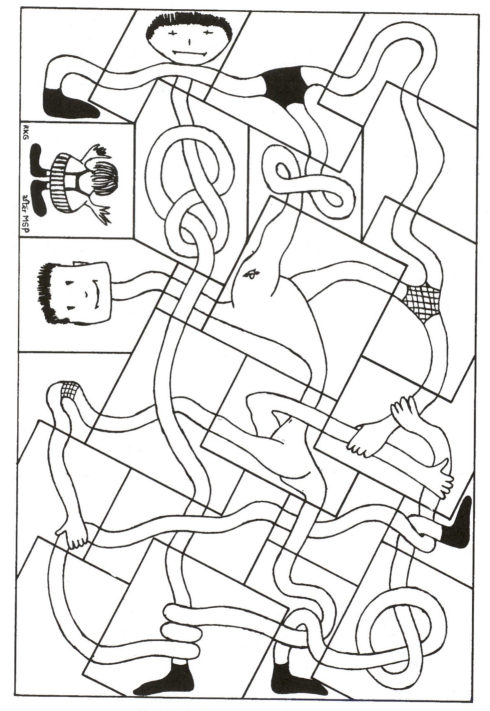

Figure 60. Paterson's Wrestling Match.

Some years ago one of us sent out a Christmas card (Fig. 59) in the form of a jigsaw puzzle based on Fig. 58(b). The assembly in Fig. 59 is *not* a solution because it contains heads connected directly to hands and necks connected directly to arms. Can you turn it into an anatomically correct solution? Figure 60 is M.S. Paterson's modification of this idea, using another shape system. You must rearrange the pieces so that each wrestler has a properly connected body consisting of one head, one torso, one pair of shorts, two arms and two legs!

Quintominal Dodecahedra

The MacMahon superdominoes with five or more sides have not received much attention, but here's a nice little problem. There are 12 different **quintominoes** if we use five different colors once each and allow turning over. Can you fit them, colors matching, onto the 12 faces of a regular dodecahedron?

The Doomsday Rule

Here's an easy way to find the day of the week for an arbitrary date in an arbitrary year. The day of the week on which the last day of February falls in any given year will be called the **doomsday** for that year. For instance, in year 1000, doomsday (Feb. 29) was a Thursday (THOUSday). Then the following dates in *any* year are all doomsdays:

$$\text{Feb } 28/29 \quad \text{Jan } 31/32$$

(the second alternative in leap years), otherwise for even months,

$$\text{Apr } 4 \quad \text{Jun } 6 \quad \text{Aug } 8 \quad \text{Oct } 10 \quad \text{Dec } 12$$

(the number of the month in the year), and for odd ones,

$$\text{Mar } 3 + 4 \quad \text{May } 5 + 4 \quad \text{Jul } 7 + 4 \quad \text{Sep } 9 - 4 \quad \text{Nov } 11 - 4$$

(add 4 for the 31-day, **long**, months; subtract 4 for 30-day, **short**, ones). Here's a summary with memos.

Jan	Feb	Mar	Apr	May	Jun	Jul	Aug	Sep	Oct	Nov	Dec
31/32	28/29	7	4	9	6	11	8	5	10	7	12
"last"	last	long 3	even 4	long 5	even 6	long 7	even 8	short 9	even 10	short 11	even 12

You should get used to finding other doomsdays in each month by changing the given one by weeks or fortnights; for example, since

$$\text{Jul 11 is a doomsday, so is Jul 4 (Independence Day),}$$

and since

$$\text{Dec 12 is a doomsday, so is Dec 26 (Boxing Day).}$$

so these are all the same day of the week (Sunday in 1937, for example).

On what day of the week was May-Day in the year 1000? May 9, and so May 2, were doomsdays (Thursdays in year 1000), so May 1 was a Wednesday.

It's easy to go wrong when adding numbers to days, so we suggest you use our mnemonics

NUN-day ONE-day TWOS-day TREBLES-day FOURS-day FIVE-day SIXER-day SE'EN-day
Sunday   Monday   Tuesday   Wednesday   Thursday   Friday   Saturday   Sunday

Let's suppose we want Michaelmas Day (Sep 29) in the year 1000: we say
    Sep 5 (short 9) and so Sep 26 are doomsdays (Thursdays—FOURS-days) so
    Sep 29 is 3 on FOURS-day = SE'EN-day (Sunday).

To find doomsday for any year in a given century, you should add to the doomsday for the century year,

    the number of *dozens* after that year,
    the *remainder* after this, and
    the number of *fours* in the remainder.

For example, for the year 1066 we say

THOUS ⎫
Thurs ⎬ day,   5 dozen,   6   and   1,
FOURS ⎭

                    (60)   (remainder)   (4's in 6)
and since   4   + 5   + 6   +   1   ≡ 2, mod 7,

doomsday in 1066 was a TWOS-day, and so the Battle of Hastings (Oct 14) was fought on a

            4 on TWOS-day = SIXER-day (Saturday).

Let's do some years in our own century, given that 1900 = Wednesday = TREBLES-day.

    Aug 4,   19————————————14
    4 off   TREBLES-day, 1 dozen, 2 (and 0) = TWOS-day (Tuesday),

    Nov 11,   19————————————18
    4 on   TREBLES-day, 1 dozen, 6 and 1 = 15-day = ONE-day (Monday).

Of course, whole weeks can be cancelled, so the parentheses in

    (4 on TREBLES)   1,   (6 and 1)

can be forgotten, making the answer immediate.

    In the Julian calendar (as instituted by Julius Caesar) each century was one day earlier than the last, and so

0	100	200	300	400	500	600
700	800	900	1000	1100	1200	1300
1400	1500	1600	1700	...		

were

    Sunday   Saturday   Friday   Thursday   Wednesday   Tuesday   Monday.

But in the modern, Gregorian, calendar (as reformed by Pope Gregory XIII)

		. . .	1500
1600	1700	1800	1900
2000	2100	2200	. . .

are

Tuesday   Sunday   Friday   Wednesday

because each century year that is *not* a multiple of 400 drops its leap day, and so is *two* days earlier than the previous one. In practice, remember that 1900 was a Wednesday, and that each step *backwards* to 1800, 1700, 1600 *adds* two days.

Thus, since Jul 4 is a doomsday,

$$\text{Jul 4,} \qquad 17\text{————————}76$$

was

exactly   Sunday, $\cancel{6}$ dozen, 4 and $\cancel{V}$ = Thursday.

Various countries adopted the Gregorian reform by omitting various days; for example,

in Italy, France and Spain,   Oct 5-14, 1582.
in Britain and the American colonies,   Sep 3-13,1752,
elsewhere, various dates between 1583 (Poland) and 1923 (Greece).

You should also remember that the start of the year has not always been Jan 1. For some time before 1066 it was Christmas Day of the previous year, and for several centuries it was Mar 25 (so called Old Style dating, which was abolished in 1752). Such things are ignored in the Doomsday Rule, but, along with varying national conventions, must be accounted for in subtle examples:

Apr 23, 1616 (England) = 2 off Friday, 1 dozen, 4 and 1 = Tuesday (Shakespeare's deathday),
Apr 23, 1616 (Spain) = 2 off Tuesday, 1 dozen, 4 and 1 = Saturday (Cervantes' deathday),
Feb 29, 1603 (England) = exactly Friday, 0 dozen, 4 and 1 = Wednesday (Whitgift's deathday).

This "1603" must obviously be 1604 (New Style). Archbishop Whitgift was Queen Elizabeth's "worthy prelate" and first chairman of the commission which eventually produced the Authorized Version of the Bible.

The ambiguous days from Jan 1 through Mar 24 in years between about 1300 and 1752 were usually written in the "double dating" convention; e.g. Queen Elizabeth's deathday was Mar. 24, 1602/3 for which we find "3 on Fri + 3" = Thursday.

When calculating a B.C. date, it's best to add a big enough multiple of 28 (or 700) years to make it into an A.D. one, remembering that there was no year 0 (1 B.C. was immediately followed by 1 A.D.). Thus, in the Julian system we add 4200 to

Oct 23, 4004 B.C., getting Oct 23, 197 A.D. (*not* 196),

and giving

1 off SIXER-day, 8 dozen, 1 (and 0) = SE'EN-day = Sunday

for the day of Creation, according to Archbishop Ussher.

Problem: 1. On what weekday is the 13th of the month most likely to fall in the Gregorian
calendar?

. . . and Easter Easily

A number of sources give more or less complicated rules for determining Easter. These usually
apply only over limited ranges and are sometimes incorrect, even in reputable works, because
they neglect the exceptions in the simple rule below.

Easter Day is defined to be the first Sunday strictly later than the **Paschal full moon**,
which is a kind of arithmetical approximation to the astronomical one. The Paschal full moon
is given by the formula

$$(\text{Apr } 19 = \text{Mar } 50) - (11G + C)_{\bmod 30}$$

except that when the formula gives

Apr 19	you should take	Apr 18

and when it gives

Apr 18 *and* $G \geq 12$,	you should take	Apr 17.

In the formula,

$$
\begin{aligned}
G(\text{the } \textbf{Golden number}) &= \text{Year}_{\bmod 19} + 1 \text{ (never forget to add the 1!)} \\
C(\text{the Century term}) &= + 3 \text{ for all Julian years} \\
& \left. \begin{array}{l} -4 \text{ for 15xx, 16xx} \\ -5 \text{ for 17xx, 18xx} \\ -6 \text{ for } \textbf{19xx}, \text{20xx, 21xx} \end{array} \right\} \text{Gregorian}
\end{aligned}
$$

The general formula for C in a Gregorian year Hxx is

$$-H + \lfloor H/4 \rfloor + \lfloor 8(H+11)/25 \rfloor .$$

The next Sunday is then easily found by the Doomsday rule. Example

$$1945 \equiv 7, \text{ mod } 19 \text{ so } G = 8 \text{ and we find for the Paschal full moon:}$$
$$\text{Mar } 50 - (88 - 6)_{\bmod 30} = \text{Mar } 50 - 22 = \text{Mar } 28.$$

Because this is a Doomsday, it's very easy to work out that it is

$$\text{"exactly Wed } (+3 + 9 + 2)\text{".}$$

Easter Day, 1945, was therefore Mar 32, April Fool's Day.

For 1981 ($\equiv 5$, mod 19) the formula gives

$$\text{Apr } 19 - (66 - 6)_{\bmod 30} = \text{Apr } 19,$$

so the Paschal full moon is

$$\text{Apr } 18 = \text{Doomsday, 1981} = \text{Saturday,}$$

so Easter Sunday, in 1981, was Apr 19.

Here is an example in the Julian system:

$$1573: \text{P.F.M.} = \text{Mar } 50 - (176 + 3)_{\bmod 30} = \text{Mar } 50 - 29 = \text{Mar } 21 = \text{Saturday},$$

so Easter Day, 1573 was Mar 22. Since this date is still in the Old Style 1572, we can say that that year contained two Easters!

You should use the Julian system even today if you want to know when the Orthodox churches celebrate Easter. Example:

$$\text{Julian P.F.M. } 1984 = \text{Apr } 19 - (99 + 3)_{\bmod 30} = \text{Apr } 7.$$

The next Doomsday is Apr 11, which is, still in the Julian system,

$$\begin{matrix} \text{Tuesday,} & \quad 7 \text{ dozen} = \text{Tuesday,} \\ (\text{Julian } 1900) & \end{matrix}$$

so that Orthodox Easter Day, 1984 is the Julian date, Apr 9. Since the Julian calendar is now 13 days out of date, this is Apr 22 in the Gregorian system.

Differences between Julian and Gregorian dates:

15xx,	16xx,	17xx,	18xx,	19xx,	20xx,	21xx,	...
10 days,	10 days,	11 days,	12 days,	13 days,	13 days,	14 days,	....

How Old is the Moon?

If you stand on the earth and watch the sun and moon going round you, you'll see that they take about $365\frac{1}{4}$ [365 · 242199] and 30 [29 · 530588 or $29\frac{5}{9}$] days to do so, on average [brackets like these contain better approximations to various numbers].

From these facts you can deduce that the number of days that have passed since the last new moon is approximately:

$$(\text{day number}) + (\text{month number}) + (\text{year number}) + (\text{century number}),$$

all reduced mod 30 [$29\frac{5}{9}$].

The **day number** is the number of the day in the month.

The **month number**

for	Jan	Feb	Mar	Apr	May	Jun	Jul	Aug	Sep	Oct	Nov	Dec
is	3	4	3	4	5	6	7	8	9	10	11	12
	[$2\frac{2}{3}$	4	$2\frac{1}{3}$	$3\frac{9}{8}$	$4\frac{4}{9}$	6	$6\frac{5}{9}$	8	$9\frac{5}{9}$	$10\frac{1}{9}$	$11\frac{5}{9}$	$11\frac{8}{9}$]

(or just remember that the rule is about $\dfrac{1}{2}$ a day late/early in the long/short odd months).

The **year number** for a year whose last two digits are congruent, modulo 19,

to	0	±1	±2	±3	±4	±5	±6	±7	±8	±9
is	0	±11	±22	±03	±14	±25	±06	±17	±28	±09
	[0	±$10\frac{8}{9}$	±$21\frac{7}{9}$	±$3\frac{1}{9}$	±14	±$24\frac{8}{9}$	±$6\frac{2}{9}$	±$17\frac{1}{9}$	±28	±$9\frac{1}{3}$]

[with an additional

	$\frac{1}{2}$	$\frac{1}{4}$	0	$-\frac{1}{4}$	$-\frac{1}{2}$
in years	$4n$ (after leap day)	$4n+1$	$4n+2$	$4n+3$	$4n+4$ (before leap day)].

The **century number** for the Gregorian centuries

	15xx	16xx	17xx	18xx	19xx	20xx	21xx	22xx	23xx	24xx
is	$16\frac{1}{3}$	12	$6\frac{2}{3}$	$1\frac{1}{3}$	-4	$-8\frac{1}{3}$	$-13\frac{2}{3}$	-19	$-24\frac{1}{3}$	$-28\frac{2}{3}$

and, for the Julian centuries

8xx	9xx	10xx	11xx	12xx	13xx	14xx	15xx	16xx	17xx
27	$22\frac{2}{3}$	$18\frac{2}{3}$	14	$9\frac{2}{3}$	$5\frac{1}{3}$	$+1$	$-3\frac{1}{3}$	$-7\frac{2}{3}$	-12

To remember these,

    the day number is easy,

    the month number also, except for Jan = 3, Feb = 4.

    the year number's tens digit is its units digit reduced, modulo 3,

    the centuries 14xx and 19xx are +1 and −4; and a short century (36524 days)
        drops back $5\frac{1}{3}$ days, while a long century (36525 days) drops back $4\frac{1}{3}$ days
        (because 1273 lunations take $36529\frac{1}{3}$ [36529 · 337] days).

Thus (using only the rough numbers) on Christmas Day, 1984, the moon will be

$$25 + 12 + (+28) - 4(\mathrm{mod}\,30) = 1 \text{ day old,}$$

since $84 \equiv +8$, mod 19 and $8 \equiv 2$, mod 3. But, applying the formula to New Year's Day 1985 we find

$$1 + 3(!) + (+09) - 4 = 9 \text{ days old}$$

despite the interval of exactly 7 days. The true motion of the moon is very complicated, and such a simple rule can only hope to give answers to within a day or so. If you're watching the moon late at night, for instance, remember that 11:00 p.m. is nearer tomorrow than today because the rule is attuned to the start of the day.

    Of course a moon's age of about

$$0, \qquad 7\frac{1}{2}, \qquad 15, \qquad 22\frac{1}{2},$$

days corresponds to

        New Moon,   First Quarter,   Full Moon,   Last Quarter.

    Those who like to keep mental track of the moon throughout a year should remember the total number for that year, e.g.

        in 1998, day number + month number -1,
        in 1999, day number + month number + 10 etc.

2000,	2001,	2002,	2003,	2004,	2005,	2006,	2007,	2008,	2009,	2010
-8,	$+3$,	± 15,	-4,	$+7$,	-12,	-1,	$+10$,	-9,	$+2$,	$+13$

Jewish New Year (Rosh Hashana)

Here's how to calculate the date of the Jewish New Year (Rosh Hashana) for and year $Y = 1900 + y$ in the range 1900 (inclusive) to 2100 (exclusive). You first compute $F = (12G)_{\bmod 19}$ and then use the formula (in which you usually ignore the fraction by which it exceeds an integer):

$$\text{Sep } 6 + \frac{3}{2}F + \frac{F+1}{18} + \frac{1}{4}y_{\bmod 4} - \frac{2y}{630}$$

However it must be postponed from any

SUN	WED	FRI	TUE if fraction $\geq \cdot633$ and if $F > 6$	MON if fraction $\geq \cdot898$ and if $F > 11$

to the following

MON	THU	SAT	THU (*not* WED)	TUE

For years outside this range, you should replace
Sep 6 by Sep 6 $\{[Y/100] - [Y/400] - 9\}$ in Gregorian years, and by Aug 24 in Julian.

$$\frac{-2y}{630} \text{ by } -\frac{2(Y-1900)-1}{630} - \frac{F}{760} = \frac{11(2100-Y)+7(F-1)}{3447360}$$

$$\cdot633 \text{ by } \frac{1367}{2160} = \cdot63287037\ldots, \quad \cdot898 \text{ by } \frac{23269}{25920} = \cdot897723765\ldots$$

(the last three replacements seldom affect the answer).

Extras

Blocks-in-a-Box

The key to this puzzle is that every piece except the three $3 \times 1 \times 1$ rods occupies as many "black" cells as "white" in every layer. The rods must therefore be arranged so as to correct the color compositions in all fifteen layers simultaneously. It turns out that there is a unique arrangement which does this. Figure 61 also shows the only three dispositions for the $2 \times 2 \times 2$ cube and $2 \times 2 \times 1$ square. With these five pieces in place, the puzzle becomes easy.

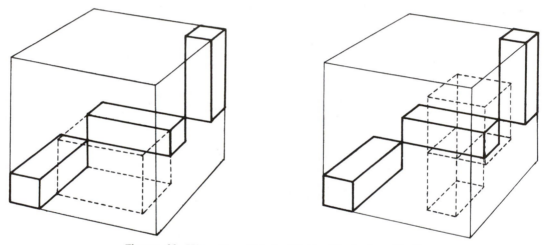

Figure 61. Were You Able to Fit the Blocks-in-a-Box?

A much harder puzzle is to pack 41 $1 \times 2 \times 4$ planks (together with 15 $1 \times 1 \times 1$ holes) into a $7 \times 7 \times 7$ box (see reference to Foregger, and to Mather, who proves that 42 planks can't be packed.)

The Somap

The Soma pieces $1 = W$, $2 = Y$ and $4 = O$, while themselves symmetrical, may appear on the surface of the cube in either the *dexter* fashion

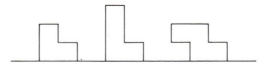

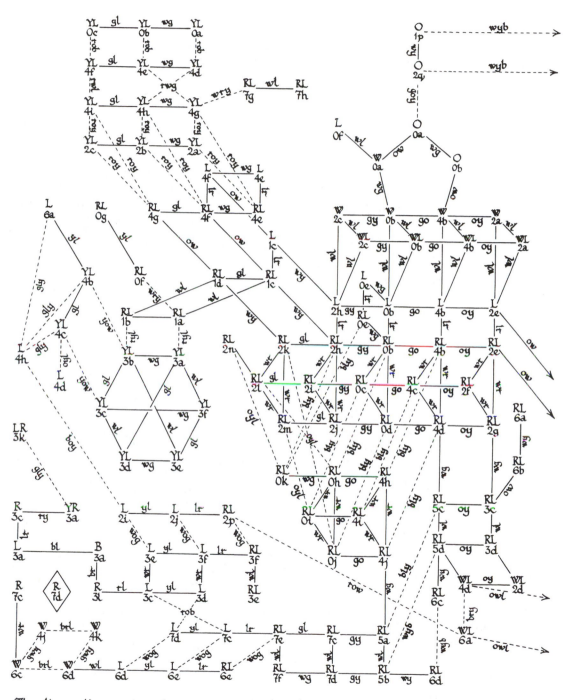

The diamond's gory secrets are seven seas away!

Figure 62. The . . .

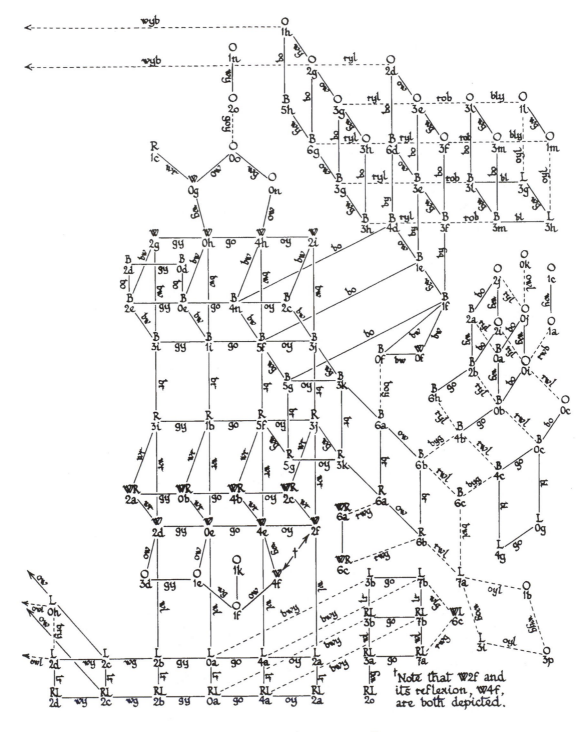

†Note that W2f and
its reflexion, W4f,
are both depicted.

. . .   .   Somap.

or the *sinister* one

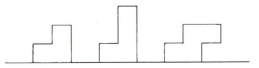

so you can tell which of these pieces are dexter by giving the sum of their numbers, which we call the **dexterity** of the solution. The symbols

$$\begin{matrix} DC & DC & DC \\ na & nb & nc \end{matrix}$$

refer to different solutions having deficient piece D, central piece C and dexterity n, a single capital letter indicating that the same piece is both deficient and central. Thus

$$\begin{matrix} RL & RL & RL & RL \\ 5a & 5b & 5c & 5d \end{matrix}$$

are four solutions in which Red is deficient, bLue is central and pieces 1 and 4 are dexter $(1 + 4 = 5)$, while

$$\begin{matrix} B & B & B \\ 6a & 6b & 6c & \dots \end{matrix}$$

are solutions in which Black is deficient *and* central while 2 and 4 are dexter.

Along with the solutions in Fig. 62, there are their reflexions whose names are found by interchanging R and L and replacing n by

$$3 - n, \qquad 6 - n, \qquad 7 - n,$$

in the cases

$$\text{O central}, \qquad \text{W central}, \qquad \text{otherwise}.$$

When two solutions are related by changing just two pieces, P and Q, this is indicated by a solid line PQ. Some three-piece changes are indicated by dashed lines in a similar way. So all that's left for you to do is to find a suitable solution which you can locate on the Somap, and this will then lead you to all the others except R7d.

Solutions to the Arithmetico-Geometric Puzzle

Figure 63 shows how we indicate layers in this puzzle by using a or α, according to orientation, for an a-high block, etc. The 21 solutions to Hoffman's puzzle are exhibited in Table 1 in this notation. When, as usual, only the middle layer is shown, another layer is separated from it by a letter S, and the remaining one is the special layer of Fig. 63. The meanings of the other letters in Table 1 are:

R: reflect the special layer across the dotted diagonal,
S: swap the two non-special layers,
S′:   swap two adjacent layers in a different direction,
T: tamper with a $2 \times 2 \times 2$ corner, not involving the special layer,
T′:   tamper with a $2 \times 2 \times 2$ corner, which does involve the special layer.

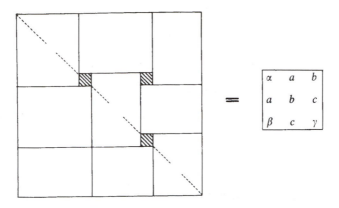

Figure 63. The Special Layer.

We'll leave it to you to work out why this gives just 21 solutions, and to verify that of these, exactly 17 have **duals**, obtained by replacing the dimensions a, b, c by c, b, a. Just one of the solutions (which?) is self-dual. This solution has the remarkable property that it can be repeatedly transformed (into rotations of itself!) by transporting either of two special faces to the opposite side.

Raphael Robinson and David Seal have found ways of combining solutions to the Arithmetico-Geometric puzzle in various dimensions to produce higher-dimensional ones. For example, if

$$a = a_1 + a_2 + a_3 \quad \text{and} \quad b = b_1 + b_2 + b_3$$

we know how to pack 27

$$a_1 \times a_2 \times a_3 \quad \text{or} \quad b_1 \times b_2 \times b_3$$

blocks into an

$$a \times a \times a \quad \text{or} \quad b \times b \times b$$

cube. The Cartesian product of these gives us a way of packing $27^2 = 729$

$$a_1 \times a_2 \times a_3 \times b_1 \times b_2 \times b_3$$

6-dimensional hyperblocks into a single

$$a \times a \times a \times b \times b \times b$$

hyperblock. But now the Cartesian product of three copies of Fig. 7 gives us a way to pack $4^3 = 64$ of these

$$
\begin{array}{cccccccccccc}
a & \times & b & \times & a & \times & b & \times & a & \times & b \\
(a+b) & \times & (a+b) & \times & (a+b) & \times & (a+b) & \times & (a+b) & \times & (a+b)
\end{array}
$$

hypercube.

In general the method combines m-dimensional and n-dimensional solutions to give an mn-dimensional one. We hope Omar will tell us how to deal with dimensions 5, 7, 11 and so on.

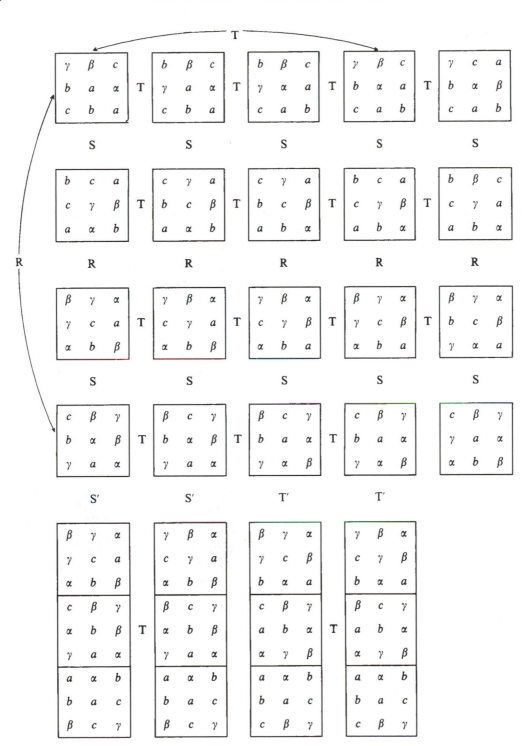

Table 1. The 21 Solutions to Hoffman's Puzzle.

...and One for "Three" Too!

0	0	2	1	1	0	1	0	0
0	0	2	1	2	1	1	2	2
2	2	1	0	1	1	0	2	2

There's only one other solution. Hint: add $x + y + z$.

Hares and Tortoises

Make the moves in this order (jumps are bold):

$$\text{H, } \mathbf{T}, \text{ T, } \mathbf{H}, \mathbf{H}, \text{ H, } \mathbf{T}, \mathbf{T}, \mathbf{T}, \text{ H, } \mathbf{H}, \mathbf{H}, \text{ T, } \mathbf{T}, \text{ H.}$$

If you move only one kind of animal for as long as you can before moving the other kind, you'll soon see how to swap 57 Hares with 57 Tortoises.

Solutions to the other coin problems (heads are **bold**) are:

Start from 012**345**;	move	01 to 67, **56** to **89** and **23** to 56;
	or	01 to 76, **23** to 98 and 56 to **65**.
Start from 01234**567**;	move	12 to 89, **45** to **12**, **78** to 45 and 01 to **78**;
	or	**67** to **98**, 01 to 76, 34 to 43 and **78** to 87.

M. Delannoy has shown that the first problem with n pairs of coins can always be solved in just n moves. However the second problem, due to Tait, requires $n + 1$ moves if $n > 4$. For some reason which we don't understand, we have always found these little problems confusing and can never remember their solutions!

The last little coin puzzle is one of the simplest examples we know of a psychological block. You notice that four coins are already in position (Fig. 64(a)), so you're reluctant either to move one of them (Fig. 64(b)) or to waste time by replacing it (Fig. 64(c)), but that's the only way to get to Fig. 64(d) in three moves. There's a four-move version in which you start with a triangle.

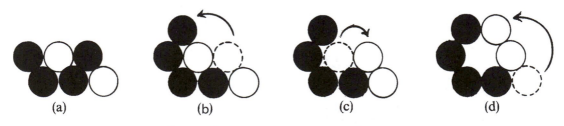

(a)   (b)   (c)   (d)

Figure 64. How to Infuriate Your Friends.

The Lucky Seven Puzzle

The Lucky Seven Puzzle has a solution in which just seven discs are slid down the bridge, alternately from the left and right sides:

$$1, 7, 2, 6, 3, 5, 4.$$

Top Face Alterations for the Hungarian Cube

We give the shortest known sequences for all permutations (Table 2) and for all combinations of flips and twists (Table 3) in the top layer. The numbers are the numbers of moves, but not counting any final top turns (U^k) which can all be saved to the end. David Seal has proved that most of these are best possible.

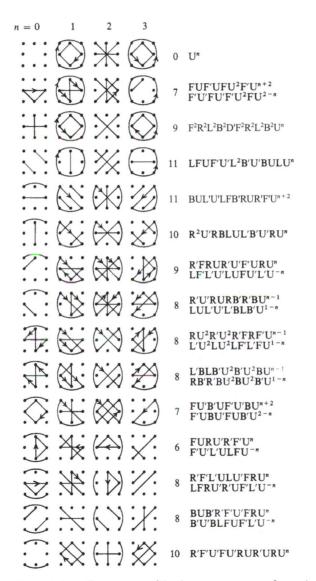

	$n=0$	1	2	3		
					0	U^n
					7	$FUF'UFU^2F'U^{n+2}$
						$F'U'FU'F'U^2FU^{2-n}$
					9	$F^2R^2L^2B^2D'F^2R^2L^2B^2U^n$
					11	$LFUF'U'L^2B'U'BULU^n$
					11	$BUL'U'LFB'RUR'F'U^{n+2}$
					10	$R^2U'RBLUL'B'U'RU^n$
					9	$R'FRUR'U'F'URU^n$
						$LF'L'U'LUFU'L'U^{-n}$
					8	$R'U'RURB'R'BU^{n-1}$
						$LUL'U'L'BLB'U^{1-n}$
					8	$RU^2R'U^2R'FRF'U^{n-1}$
						$L'U^2LU^2LF'L'FU^{1-n}$
					8	$L'BLB'U^2B'U^2BU^{n-1}$
						$RB'R'BU^2BU^2B'U^{1-n}$
					7	$FU'B'UF'U'BU^{n+2}$
						$F'UBU'FUB'U^{2-n}$
					6	$FURU'R'F'U^n$
						$F'U'L'ULFU^{-n}$
					8	$R'F'L'ULU'FRU^n$
						$LFRU'R'UF'L'U^{-n}$
					8	$BUB'R'F'U'FRU^n$
						$B'U'BLFUF'L'U^{-n}$
					10	$R'F'U'FU'RUR'URU^n$

Table 2. Top Layer Permutation Sequences. (the lower sequence of a pair refers to the left-right reflected picture)

Pattern	Moves	Sequence
`• • •` `• • •` `• • •`	0	no moves required

`• • •`
`•   e`
`• e •`
	12	F'U'F²DRUR'D'U'F²U²FU'

`• e •`
`•   •`
`• e •`
	13	LF'UL'FB'UR'FU'RF'BU'

`• e •`
`e   e`
`• e •`
	13	L²F²L²U²R'LFL'RU²L²F²L²U

`• • a`
`•   •`
`• • c`
	12	R'U'LU²R'F²RF²U'RU²L'U²
		RU²L'UB²L'B²LU²R'ULU²

`• • a`
`•   e`
`• e c`
	13	RU'LU²R²F'U'FUR²U²R'L'U
		BFU²F²U'L'ULF²U²B'UF'U'

`• e a`
`•   e`
`• • c`
	13	R'UL'U²R²BUB'U'R²U²RLU'
		F'B'U²B²ULU'L'B²U²FU'BU

`e • a`
`•   •`
`• e c`
	14	F²DF'UFD'FL'U'LUF²U²F'U
		LU²L²U'B'UBL'DL'U'LD'L'U'

`• e a`
`e   •`
`• • c`
	14	B²D'BU'B'DB'LUL'U'B²U²BU'
		L'U²L²UFU'FLD'LUL'DL²U

`• • a`
`•   •`
`• c c`
	14	RUR²F²D'R²BL'B'R²DF'RF'U²
		BL'BD'L²FRF'L²DB²L²U'L'U²

`• • a`
`•   •`
`• • c`
	14	LF'D'L'BL'B²U²L'BDF'L²F²U²
		B²R²BDF'RU²F²RF'RDBR'U²

`• e a`
`e   e`
`• e c`
	15	BU²BR²FD²FLFL²F²D²F'R²B²U
		R²F²LD²L²B²L'B'L'D²L'F'R'U²R'U'

`a • •`
`•   •`
`• • c`
	11	R'BD²B'RU²R'BD²B'RU²

`a • •`
`•   e`
`• e c`
	14	B'UFU'BU²F²L'U'LUF²U²F'
		LU'R'UL'U²R²BUB'U'R²U²R

`a • •`
`e   •`
`• e c`
	13	L'UBL'D'BD²R²D'B²L²UF²U²
		B²U'R²F²DL²D²F'DRF'U'RU²

`a e •`
`•   •`
`• e c`
	13	R'B²F'L'DF'L²FD'LFB²RU²

`a • •`
`e   e`
`• • c`
	13	BR²LFD'LF²L'DF'L'R²B'U²

`a e •`
`e   e`
`• e c`
	15	F'L²D'B²D'R²B'R'F'R²D'F²D'L²B'U

`c • c`
`•   •`
`• • c`
	12	RU'L'UDB²D'R²U²LU'RU'
		B'UF'U²B²DL²U'D'FUB'U'

`c • c`
`•   e`
`• e c`
	11	LR'U'R²B'R'B²U'B'U²L'
		LU²BUB²RBR²URL'

`c e c`
`e   •`
`• • c`
	11	F'U²R'U'R²B'R'B²U'B'F
		F'BUB²RBR²URU²F

`c • c`
`e   •`
`• e c`
	14	F²R²FU'F²R'F'R²U'R'U²F²R²F²
		F²R²F²U²RUR²FRF²UF'R²F²

`c e c`
`•   •`
`• • c`
	8	B'U'B²L'B'L²U'L'U²
		RUR²FRF²UFU²

`c e c`
`•   •`
`• e c`
	12	BLU²B'U'B²L'B'L²U'L²B'
		BL²UL²BLB²UBU²L'B'

`c • c`
`e   e`
`• • c`
	12	R'F²U'F'R'F'R²U'R'U²FR
		R'F'U²RUR²FRF²UF²R

`c e c`
`e   e`
`• e c`
	12	L'U'B'UBLBLUL'U'B'U²
		FURU'R'F'R'F'U'FURU²

`a • c`
`•   •`
`a • c`
	14	R²B²R²U'RL'DL'U²LD'L'U²R'U

`a • c`
`•   e`
`a e c`
	12	LUFU'F'L'R'U'F'UFR
		R'F'U'FURLFUF'U'L'

`a e c`
`•   e`
`a • c`
	12	RBUB'U'R'L'B'U'BUL
		L'U'B'UBLRUBU'B'R'

`a e c`
`•   •`
`a e c`
	14	F'U'F²UL'U'RLU'R'UF²U'F'U²

`a • c`
`e   e`
`a • c`
	14	R'U'F'L'R'U'F'UFRLUFR

`a e c`
`e   e`
`a e c`
	13	LU²F'U'F²LF²L'F²UFU²L'U²

`c • a`
`•   •`
`a • c`
	15	R'U²RU²R²B'D'R'FR²F'DBU'R'U'

`c • a`
`•   e`
`a e c`
	15	R'F²U'D'L'F'LFDULFL'FR
		R'F'LF'L'U'D'F'L'FLDUF²R

`c e a`
`•   •`
`a e c`
	14	R'UB'R²U'RUR²BU²R²R'B'R'BU²
		F'LFL²U²F²L²U'L'UL²FU'LU²

`c e a`
`e   e`
`a e c`
	14	B'D²FU²RF'D²BLU²FU²F'LU

Table 3. Top Layer Flip and Twist Sequences. (the lower sequence of a pair refers to the picture with a and c swapped)

The Century Puzzle

The Century Puzzle is so called because it takes exactly 100 moves, and the Century-and-a-Half takes 151 according to the official rules, but since the first and last are only half-moves, we can obviously count it as 150 whole moves. You can see solutions to both puzzles in Fig. 65, and by turning the book upside-down you'll see the only other 100-move solution to the Century.

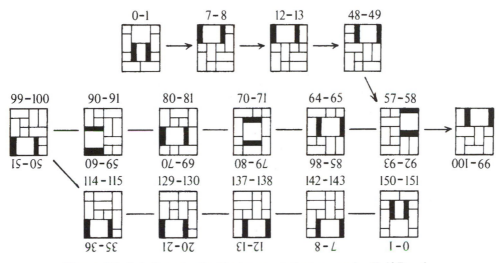

Figure 65. Solutions to the Century and Century-and-a-Half Puzzles.

Adams's Amazing Magic Hexagon

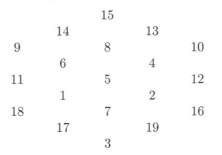

In Martin Gardner's *Sixth Book of Mathematical Games* you can read the remarkable history of Clifford W. Adams's discovery and of Charles W. Trigg's uniqueness proof. It's easy to see that a diameter d magic hexagon uses the numbers from 1 to $(3d^2 + 1)/4$, which add to

$$\frac{1}{2}\left(\frac{3d^2+1}{4}\right)\left(\frac{3d^2+5}{4}\right) = \frac{1}{32}\left(9d^4 + 18d^2 + 5\right),$$

so that each of its d columns must add to

$$\frac{1}{32}\left(9d^3 + 18d + \frac{5}{d}\right)$$

which can only be an integer if d divides 5. Frank Tapson has discovered that one William Radcliffe had already found the 'Adams Hexagon' and registered it at Stationers Hall, London, in 1896. We thank Victor Meally for this information.

Flags of the Allies Solution

If you use the O'Beirne method you will find the two pairs of 5-circuits

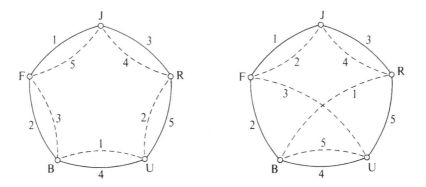

which lead to the solutions shown in Fig 49 and Fig. 66.

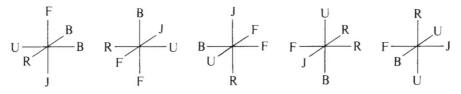

Figure 66. The Other Solution to the Flags of the Allies Problems.

All Hexiamond Solutions Found

In May 1996 Marc Paulhus wrote a program that used only a few days of computer time to find all solutions to O'Beirne's Hexiamond puzzle. Independently, in 1999, Donald Knuth also ran a program (in 5 hours and 21 minutes on a 450 MHz Pentium II) which found one more solution than Paulhus had found. Paulhus reran his program and this time really found them all. The initial run must have contained a machine error!

The numbers of solutions, classified according to how far out the Hexagon appears: A in a corner, B in a side, C, ..., F, G in the centre, are

A 75490   B 15717   C 6675   D 7549   E 11447   F 5727   G 1914

with a total of 124519 solutions, of which Figure 52 is indeed the most symmetrical.

The Three Quintominal Dodecahedra

The three quintominal dodecahedra should be recoverable from

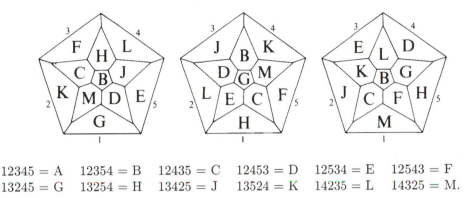

$$12345 = A \quad 12354 = B \quad 12435 = C \quad 12453 = D \quad 12534 = E \quad 12543 = F$$
$$13245 = G \quad 13254 = H \quad 13425 = J \quad 13524 = K \quad 14235 = L \quad 14325 = M.$$

Answer to Exercise for Experts

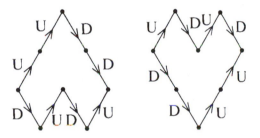

We have a rather complicated proof that n^2 copies of hexiamond A can be used to replicate A on a larger scale only if $n \equiv 0$ or ±1 mod 6. Our proof establishes that these are the only values of n for which the relations (look at the foot of the previous page)

$$U^2D^2 = DUDU \text{ and } D^2U^2 = UDUD$$

imply

$$U^{2n}D^{2n} = D^nU^nD^nU^n.$$

We've also shown that none of the usual kinds of coloring argument excludes other values of n.

Where Do the Black Edges of MacMahon Squares Go?

Round the outside, of course, but there are six more inside. These can be arranged in 20 different ways. In the first two the ladder is in the *third* column, otherwise it's in the second. The last row of Fig. 67 contains $6 + 6 + 2$ arrangements: the dotted lines are alternative positions for the sixth black edge.

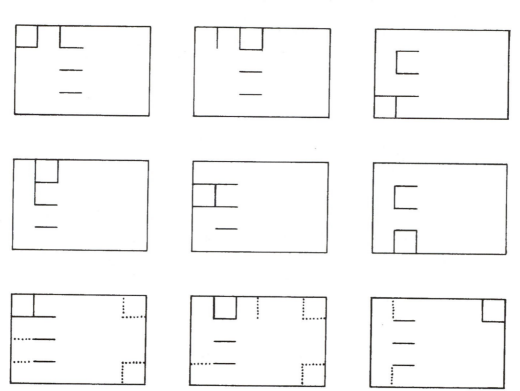

Figure 67. The Twenty Black Edge Arrangements for MacMahon Squares.

A Doomsday Answer

A tedious enumeration shows that in the 400 years of the Gregorian cycle Doomsday is

	Sun	Mon	Tue	Wed	Thu	Fri	Sat	
for	43	43	43	43	44	43	44	ordinary years
and	13	15	13	15	13	14	14	leap years

From this you can work out that the 13th day falls on

	Sun	Mon	Tue	Wed	Thu	Fri	Sat	
in	687	685	685	687	684	688	684	months,

verifying B.H. Brown's assertion that the 13th of a month is just a little bit more likely to be a Friday than any other day of the week!

References and Further Reading

S. N. Afriat, The Ring of Linked Rings, Duckworth, London, 1982 (Chinese rings).

W.S. Andrews, "Magic Squares and Cubes", Open Court, 1917, reprinted Dover, 1960.

A.K. Austin, The 14-15 puzzle. Note 63.5, Math. Gaz. 63 (1979) 45-46.

W.W. Rouse Ball and H.S.M. Coxeter , "Mathematical Recreations and Essays", 12th edn. University of Toronto Press, 1974, pp. 26-27 (calendar problems), pp. 116-118 (shunting problems), p. 121 (sliding coins), pp. 193-221 (magic squares), pp. 312-322 (Fifteen Puzzle, Tower of Hanoï', Chinese rings). See early editions for Tangrams. Pages 141-144 on equilateral zonohedra and the references there, are related to Schoen's Cyclotome puzzles.

C.J. Bouwkamp , Catalogue of solutions of the rectangular $2 \times 5 \times 6$ solid pentomino problem, Nederl. Akad. Wet. Proc. Ser. A, 81 (1978) 177-186; Zbl. 384.42011.

Bro. Alfred Brousseau , Tower of Hanoi' with more pegs, J. Recreational Math., 8 (1975-76) 169-176.

B.H. Brown , Problem E36, Amer. Math. Monthly, 40 (1933) 295 (calendar).

T.A. Brown , A note on "Instant Insanity", Math. Mag. 41 (1968) 68.

Cardan , "De Subtilitate", book xv, para 2; ed. Sponius vol. III, p. 587 (Chinese rings).

F. de Carteblanche , The coloured cubes problem, Eureka, 9 (1947) 9-11 (Tantalizer).

T.R. Dawson and W.E. Lester , A notation for dissection problems. Fairy Chess Review, 3 (Apr 1937) 5, 46-47 (polyominoes).

M. Delannoy , La Nature, June 1887, p. 10 (sliding coins).

A.P. Domoryad , "Mathematical Games and Pastimes", Pergamon, 1963, pp. 71-74 (Chinese rings); pp. 75-76 (Tower of Hanoï'); pp. 79-85 (Fifteen puzzle); pp. 97-104 (magic squares); pp. 127-128 (sliding coins); pp. 142-144 (cf. Schoen's puzzle).

Henry Ernest Dudeney , "The Canterbury Puzzles", Nelson, London, 1907, 1919 reprinted Dover, 1958, No. 74 The Broken Chessboard, pp. 119-121, 220-221 (pentominoes).

Henry Ernest Dudeney , "536 Puzzles and Curious Problems", ed. Martin Gardner, Chas. Scribner's Sons, New York 1967, No. 383 The six pennies, pp. 138, 343; No. 377 Black and white, pp. 135, 340; No. 516 A calendar puzzle, pp. 212, 409-410; No. 528 A leap year puzzle, pp. 217, 413.

T.H. Foregger , Problem E2524, Amer Math. Monthly, 82 (1975) 300; solution Michael Mather, 83 (1976) 741-742.

Martin Gardner , Mathematical Games, Sci. Amer., 196 #5 (May 1957) (Tower of Hanoï); 197 #6 (Dec. 1957) 203 #5(Nov 1960) 186-194, 207 #5(Nov 1962) 151-159 (Polyominoes); 199 #3 (Sept 1958) 182-188 (Soma); 210 #2 (Feb 1964) 122-126, 222 #2 (Feb 1970) (Fifteen puzzle. Sliding block puzzles); 210 #3 (Mar 1964) 126-127 (Magic squares); 219 #4 (Oct 1968) 120-125 (MacMahon triangles; Conway's "three" puzzle); 223 #6 (Dec 1970) 110-114 (non-transitive dice; quintominal dodecahedra).

Martin Gardner , "The Scientific American Book of Mathematical Puzzles and Diversions", Simon and Schuster, New York 1959, pp. 15-22 (magic squares); pp. 55-62 (Tower of Hanoï); pp. 88 (Fifteen puzzle); pp. 124-140 (polyominoes).

Martin Gardner , "The 2nd Scientific American Book of Mathematical Puzzles and Diversions", Simon and Schuster, New York, 1961, pp. 55-56, 59 (sliding pennies); pp. 65-77 (Soma); pp. 130-140 (Magic squares); pp. 214-215, 218-219 (another Solitaire-like puzzle).

Martin Gardner , "Sixth Book of Mathematical Games from Scientific American", Chas. Scribner's Sons, New York, 1971, pp. 23-24 (magic hexagon); pp. 64-70 (sliding block puzzles); pp. 173-182 (polyiamonds).

Martin Gardner , "Mathematical Puzzles of Sam Loyd", Dover, New York 1959, No. 73 pp. 70, 146-147.

S.W. Golomb , Checkerboards and polyominoes, Amer. Math. Monthly, 61 (1954) 675-682.

S.W. Golomb, The general theory of polyominoes. Recreational Math. Mag. 4 (Aug 1961) 3-12; 5 (Oct 1961) 3-12; 6 (Dec 1961) 3-20; 8 (Apr 1962) 7-16.

S.W. Golomb , "Polyominoes", Chas. Scribner's Sons, New York, 1965.

A.P. Grecos and R.W. Gibberd, A diagrammatic solution to "Instant Insanity" problem. Math. Mag. 44 (1971)71.

L. Gros, "Théorie du Baguenodier", Lyons, 1872 (Chinese rings).

P. Michael Grundy, Richard S. Scorer and Cedric A. B. Smith, Some binary games, *Math. Gaz.*, 28 (1944) 96–103.

L.J. Guibas and A.M. Odlyzko, Periods in strings, J. Combin. Theory Ser. A 30 (1981) 19-42.

L.J. Guibas and A.M. Odlyzko, String overlaps, pattern matching and non-transitive games, J. Combin. Theory Ser. A 30 (1981) 183-208.

Richard K. Guy, O'Beirne's Hexiamond, in Elwyn Berlekamp and Tom Rodgers, *The Mathemagician and Pied Puzzler*, A Collection in Tribute to Martin Gardner, A K Peters, Ltd., 1999, 85–96.

Bela Hajtman, On coverings of generalized checkerboards, I, Magyar Tud. Akad. Math. Kutato Int. Koz, 7(1962)53-71.

Sir Paul Harvey, "The Oxford Companion to English Literature", 4th ed., Oxford, 1967, Appendix III, The Calendar.

C.B. and Jenifer Haselgrove, A computer program for pentominoes. Eureka, 23 (1960) 16-18.

Douglas R. Hofstadter, Metamagical Themas: The Magic Cube's cubies are twiddled by cubists and solved by cubemeisters, Sci. Amer. 244 #3 (Mar. 1981) 20-39.

J.A. Hunter and Joseph S. Madachy, "Mathematical Diversions", Van Nostrand, New York, 1963, Chapter 8, Fun with Shapes, pp. 77-89.

Maurice Kraitchik, "Mathematical Recreations", George Alien and Unwin, 1943, pp. 89-93 (Chinese rings, Tower of Hanoï); pp. 109-116 (calendar); pp. 142-192 (magic squares); pp. 222-226 (shunting puzzles); pp. 302-308 (Fifteen puzzle).

Kay P. Litchfield , A 2 × 2 × 1 solution to "Instant Insanity", Pi Mu Epsilon J. 5 (1972) 334-337.

E. Lucas, "Récréations Mathematiques", Gauthier-Villars 1882-94; Blanchard Paris, 1960.

Major P.A. MacMahon, "New Mathematical Pastimes", Cambridge University Press, 1921.

Kersten Meier, Restoring the Rubik's Cube. A manual for beginners, an improved translation of "Puzzles-pass mit dem Rubik's Cube" 4c Hulme, Escondido Village, Stanford CA 94305 USA or Henning-Storm-Str. 5, 221 Itzehoe, W. Germany, 1981:01:20.

J.C.P. Miller, Pentominoes, Eureka, 23 (1960) 13-16. T.H. O'Beirne, "Puzzles and Paradoxes", Oxford University Press, London, 1965, pp. 112-129 (Tantalizer); pp. 168-184 (Easter).

T.H. O'Beirne, Puzzles and Paradoxes, in *New Scientist*, 258 (61:10:26) 260-261; 259 (61:11:02) 316-317; 260 (61:11:9) 379-380; 266 (61:12:21) 751-752; 270 (62:01:18) 158-159.

Ozanam, "Recreations", 1723, vol. IV 439, (Chinese rings).

De Parville, La Nature, Paris, 1884, part i, 285-286 (Tower of Hanoï).

B.D. Price, Pyramid Patience, Eureka, 8 (1944) 5-7.

R.C. Read, Contributions to the cell growth problem, Canad. J. Math. 14 (1962) 1-20 (polyominoes).

J.E. Reeve and J.A. Tyrrell, Maestro puzzles. Math. Gaz. 45 (1961) 97-99 (polyominoes).

Raphael Robinson, Solution to problem E36, Amer. Math. Monthly, 40 (1933) 607.

Barkley Rosser and R.J. Walker, On the transformation group for diabolic magic squares of order four, Bull. Amer. Math. Soc. 44 (1938) 416-420.

Barkley Rosser and R.J. Walker, The algebraic theory of diabolic magic squares, Duke Math. J. 5 (1939) 705-728.

T. Roth, The Tower of Bramah revisited, J. Recreational Math. 7 (1974) 116-119.

Wolfgang Alexander Schocken, "The Calculated Confusion of Calendars", Vantage Press, New York, 1976.

Leslie E. Shader, Cleopatra's pyramid. Math. Mag. 51 (1978) 57-60 (Tantalizer variant).

David Singmaster, Notes on the 'magic cube', Dept. Math. Sci. & Comput. Polytech. of S. Bank, London SE AA, England, 1979, 5th edition, 1980, £2-00 or $5-00.

W. Stead, Dissection, Fairy Chess Review, 9 (Dec 1954) 2-4 (polyominoes).

James Ussher, "Annales Veteris Testamenti", Vol. 8, Dublin ed. 1864, p. 13 ("beginning of night leading into Oct. 23").

Joan Vandeventer, Instant Insanity, in "The Many Facets of Graph Theory", Springer Lecture Notes 110 (1969) 283-286.

Wallis, "Algebra", latin edition 1693. Opera, Vol. II, Chap. cxi, pp. 472-478 (Chinese rings).

Harold Watkins, "Time counts; the story of the calendar", Neville Spearman, London, 1954.

Richard M. Wilson, Graph puzzles, homotopy and the alternating group, J. Combin. Theory Ser. B 16 (1974) 86-96.

-25-

What is Life?

Life's not always as simple as mathematics, Abraham!
　　Mrs. Abraham Fraenkel.
Life's too important a matter to be taken seriously.
　　Oscar Wilde.

... in real life mistakes are likely to be irrevocable.
Computer simulation, however, makes it economically practical
to make mistakes on purpose. If you are astute, therefore,
you can learn much more than they cost. Furthermore, if you
are at all discreet, no one but you need ever know you made
a mistake.
　　John McLeod and John Osborn, *Natural Automata*
　　and *Useful Simulations*, Macmillan, 1966.

Most of this book has been about two-player games, and our last two chapters were about one player games. Now we're going to talk about a no-player game, the **Game of Life**! Our younger readers won't have learned much about Life, so we'd better tell you some of the facts.

Life is a "game" played on an infinite squared board. At any time some of the cells will be **live** and others **dead**. Which cells are live at time 0 is up to you! But then you've nothing else to do, because the state at any later time follows inexorably from the previous one by the rules of the game:

BIRTH. A cell that's *dead* at time t becomes *live* at $t + 1$ only if *exactly three* of its eight neighbors were live at t.

DEATH by overcrowding. A cell that's live at t and has four or more of its eight neighbors live at t will be dead by time $t + 1$.

DEATH by exposure. A live cell that has only one live neighbor, or none at all, at time t, will also be dead at $t + 1$.

These are the only causes of death, so we can take a more positive viewpoint and describe instead the rule for

SURVIVAL. A cell that was live at time t will remain live at $t + 1$ if and only if it had just 2 or 3 live neighbors at time t.

Just 3 for BIRTH
2 or 3 for SURVIVAL

A fairly typical Life History is shown in Fig. 1. We chose a simple line of five live cells for our generation 0. In the figures a circle denotes a live cell .

Which of these will survive to generation 1? The two end cells have just one neighbor each and so will die of exposure. But the three inner ones have two living neighbors and so will survive. That's why we've filled in those circles.

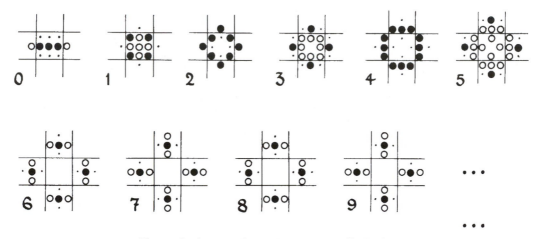

Figure 1. A Line of Five Becomes Traffic Lights.

What about births at time 1? There are three cells on either side of the line that are dead at time 0, but have exactly three live neighbors, so will come to life at time 1. We've shown these prospective births by dots in the figure.

So at time 1 the configuration will be a solid 3 x 3 square. Let's briefly follow its later progress.

Time 1-2: The corners will survive, having 3 neighbors each, but everything else will die of overcrowding. There will be 4 births, one by the middle of each side.

2-3: We see a ring in which each live cell has 2 neighbors so everything survives; there are 4 births inside.

3-4: Massive overcrowding kills off all except the 4 outer cells, but neighbors of these are born to form:

4-5: another survival ring with 8 happy events about to take place.

5-6: More overcrowding again leaves just 4 survivors. This time the neighboring births form:

6-7: four separated lines of 3, called **Blinkers**, which will never interact again.

7-8-9-10-... At each generation the tips of the Blinkers die of exposure but the births on each side reform the line in a perpendicular direction.

The configuration will therefore oscillate with period two forever. The final pair of configurations is sufficiently common to deserve a name. We call them **Traffic Lights**.

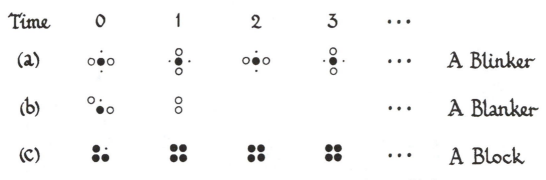

Figure 2. If Three Survive, They'll Make a Blinker or a Block.

The Blinker is also quite common on its own (Fig. 2a). Most other starting configurations of three live cells will blank out completely in two moves (Fig. 2(b)). But if you start with three of the four cells of a 2 × 2 block, the fourth cell will be born and then the **Block** will be stable (Fig. 2(c)) because each cell is neighbored by the three others.

Still Life

It's easy to find other stable configurations. The commonest such **Still Life** can be seen in Fig. 3 along with their traditional names. The simple cases are usually loops in which each live cell has two or three neighbors according to local curvature, but the exact shape of the loop is important for effective birth control.

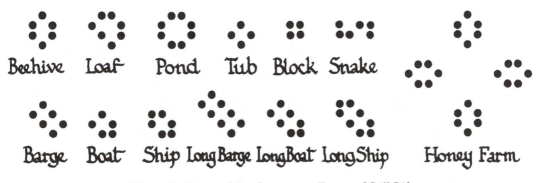

Figure 3. Some of the Commoner Forms of Still Life.

Life Cycles

The blinker is the simplest example of a configuration whose life history repeats itself with period > 1. Lifenthusiasts (a word due to Robert T. Wainwright) have found many other such configurations, a number of which are shown in Figs. 4 to 8.

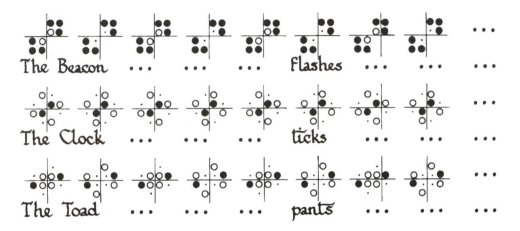

Figure 4. Three Life Cycles with Period Two.

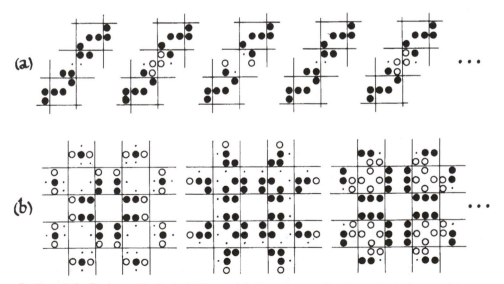

Figure 5. Two Life Cycles with Period Three, (a) Two Eaters Gnash at Each Other, (b) The Cambridge Pulsar CP 48-56-72.

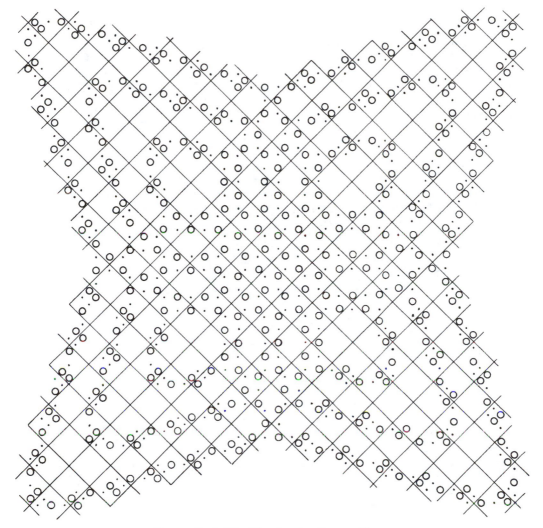

Figure 6. A Flip-Flop by the Gosper Group.

The Glider and Other Space Ships

When we first tracked the r-pentomino (you'll hear about that soon) some guy suddenly said, "Come over here, there's a piece that's walking!" We came over and found Fig. 9.

You'll see that generation 4 is just like generation 0 but moved one diagonal place, so that the configuration will steadily move across the plane. Because the arrangements at times 2, 6, 10, ... are related to those at times 0, 4, 8, 12, ... by the symmetry that geometers call a *glide reflexion*, we call this creature the **glider** . But when you see Life played at the right

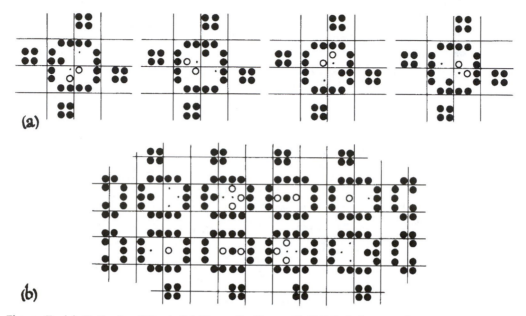

(a)

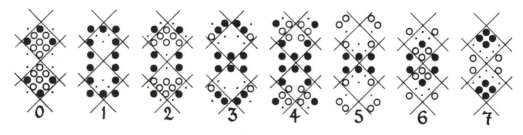

(b)

Figure 7. (a) Catherine Wheel, (b) Hertz Oscillator. Still Life Induction Coils Keep Field Stable.

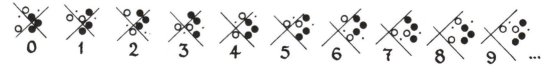

Figure 8.

Figure 9. The Glider Moves One Square Diagonally Each Four Generations.

speed by a computer on a visual display, you'll see that the glider walks quite seductively, wagging its tail behind it. We'll see quite a lot of the glider in this chapter.

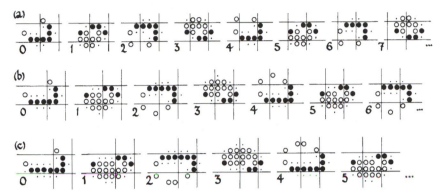

Figure 10. (a) Lightweight, (b) Middleweight, (c) Heavyweight Spaceships.

It was at just such a visual computer display that one of us first noticed the **spaceship** of Fig. 10(a) (and was very lucky to be able to stop the machine just before it would have crashed into another configuration). This **lightweight spaceship** immediately generalizes to the **middleweight** and **heavyweight** ones (Figs. 10(b) and (c)) but longer versions turn out to be unstable. It was later discovered, however, that arbitrarily long spaceships can still travel provided they are suitably escorted by small ones (Fig. 11). All the spaceships, as drawn, move Eastwards.

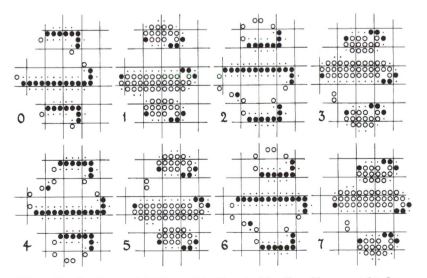

Figure 11. An Overweight Spaceship Escorted by Two Heavyweight Ones.

The Unpredictability of Life

Is there some way to foretell the destiny of a Life pattern? Will it eventually fade away completely? Or become static? Oscillate? Travel across the plane, or maybe expand indefinitely? Let's look at what should be a very simple starting configuration—a straight line of n live cells

$n = 1$ or 2 fades immediately,

$n = 3$ is the Blinker;

$n = 4$ becomes a Beehive at time 2,

$n = 5$ gave Traffic Lights (Fig. 1) at time 6,

$n = 6$ fades at $t = 12$,

$n = 7$ makes a beautifully symmetric display before terminating in the **Honey Farm** (Fig. 3) at $t = 14$;

$n = 8$ gives 4 blocks and 4 beehives,

$n = 9$ makes two sets of Traffic Lights,

$n = 10$ turns into the pentadecathlon, with a life cycle of 15,

$n = 11$ becomes two blinkers,

$n = 12$ makes two beehives,

$n = 13$ turns into two blinkers,

$n = 14$
and   }   vanish completely,
$n = 15$

$n = 16$ makes a big set of Traffic Lights with 8 blinkers,

$n = 17$ becomes 4 blocks,

$n = 18$
and   }   fade away entirely,
$n = 19$

$n = 20$ makes just 2 blocks,

and so on.

What's the general pattern? Even when we follow the configurations which start with a very small number of cells, it's not easy to see what goes on. There are 12 edge-connected regions with 5 cells (S.W. Golomb calls them pentominoes). Here are their histories:

Figure	Our mnemonic	Destiny

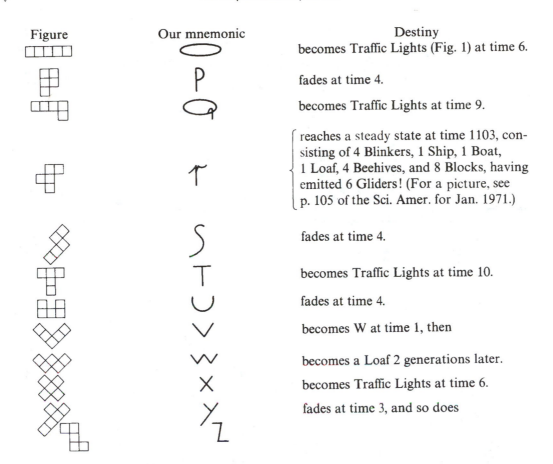

becomes Traffic Lights (Fig. 1) at time 6.

fades at time 4.

becomes Traffic Lights at time 9.

reaches a steady state at time 1103, consisting of 4 Blinkers, 1 Ship, 1 Boat, 1 Loaf, 4 Beehives, and 8 Blocks, having emitted 6 Gliders! (For a picture, see p. 105 of the Sci. Amer. for Jan. 1971.)

fades at time 4.

becomes Traffic Lights at time 10.

fades at time 4.

becomes Traffic Lights at time 6.

becomes W at time 1, then

becomes a Loaf 2 generations later.

becomes Traffic Lights at time 6.

fades at time 3, and so does

Once again, it doesn't seem easy to detect any general rule.

Here, in Figs. 12 and 13 are some other configurations with specially interesting Life Histories, for you to try your skill with.

Can the population of a Life configuration grow without limit? Yes! The $50.00 prize that one of us offered for settling this question was won in November 1970 by a group at M.I.T. headed by R.W. Gosper. Gosper's ingenious **glider gun** (Fig. 14) emits a new glider every 30 generations. Fortunately it was just what we wanted to complete our proof that

Life is really
unpredictable!

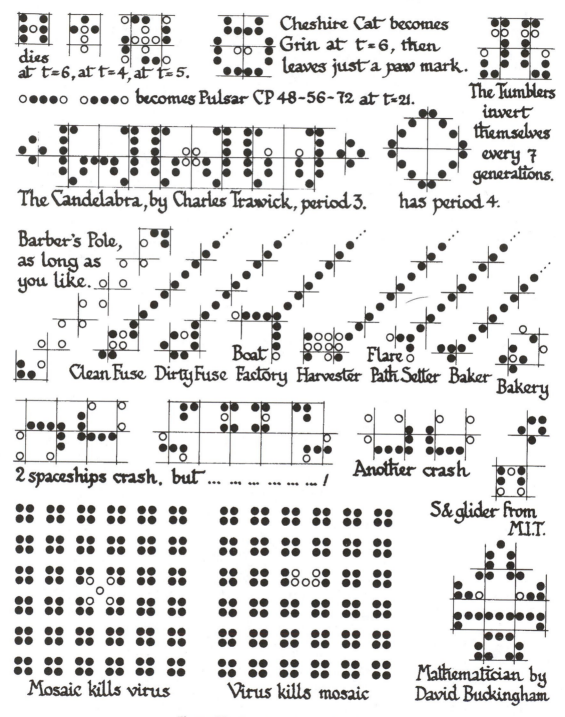

Figure 12. Exercises for the Reader.

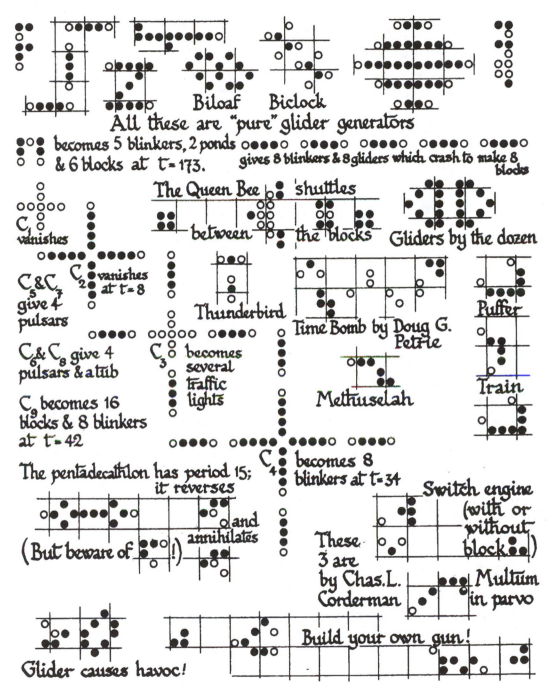

All these are "pure" glider generators

becomes 5 blinkers, 2 ponds & 6 blocks at $t = 173$.

gives 8 blinkers & 8 gliders which crash to make 8 blocks

Biloaf   Biclock

The Queen Bee 'shuttles

C_1 vanishes

between   the blocks   Gliders by the dozen

C_5 & C_7 give 4 pulsars

C_2 vanishes at $t = 8$

Thunderbird

C_6 & C_8 give 4 pulsars & a tub

C_3 becomes several traffic lights

Time Bomb by Doug G. Petrie

Methuselah

Puffer

Train

C_9 becomes 16 blocks & 8 blinkers at $t = 42$

C_4 becomes 8 blinkers at $t = 34$

The pentadecathlon has period 15; it reverses

(But beware of   !)

and annihilates

These 3 are by Chas. L. Corderman

Switch engine (with or without block   )

Multum in parvo

Glider causes havoc!

Build your own gun!

Figure 13. Mainly for Computer Buffs.

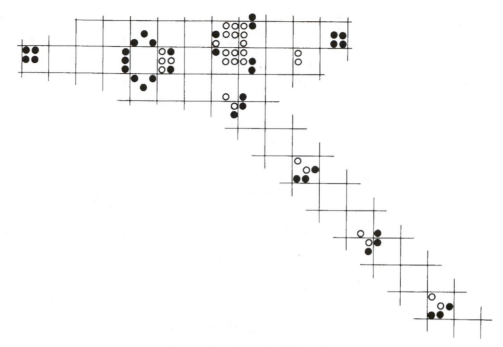

Figure 14. Gosper's Glider Gun.

Gardens of Eden

There are Life configurations that can only arise as the initial state, because they have no ancestors!

We'll prove that if n is sufficiently large, there is some configuration within a $5n - 2$ by $5n - 2$ square that has no parent. It will suffice to examine that part of a prospective parent that lies in the surrounding $5n \times 5n$ square (Fig. 15). If any one of the component 5×5 squares is empty, it can be replaced as in Fig. 15(b) without affecting subsequent generations. So we need consider only

$$(2^{25} - 1)^{n^2} = 2^{24 \cdot 999999957004337 \ldots n^2} \text{ of the } 2^{25n^2}$$

configurations in the $5n$ by $5n$ square. However, there are exactly

$$5^{(5n-2)^2} = 2^{25n^2 - 20n + 4}$$

possible configurations in the $5n - 2$ by $5n - 2$ square, so that if

$$24 \cdot 999999957004337 \ldots n^2 < 25n^2 - 20n + 4,$$

one of these will have no parent! We calculate that this happens for $n = 465163200$, so that there is a Garden of Eden configuration that will fit comfortably inside a

2325816000 by 2325816000 square!

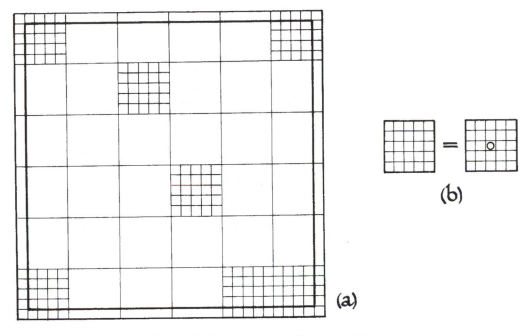

Figure 15. Location of the Garden of Eden.

This type of argument was first used by E.F. Moore in a more general context. More careful counting in the Life case has brought the size down to 1400 by 1400. However, using completely different ideas and many hours of computer time the M.I.T. group managed to produce an explicit example (Fig. 16).

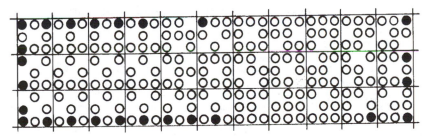

Figure 16. An Orphan Found by Roger Banks, Mike Beeler, Rich Schroeppel, Steve Ward, et al.

Life's Problems are Hard!

The questions we posed about the ultimate destiny of Life configurations may not seem very mathematical. After all, Life's but a game! Surely there aren't any difficult mathematical problems there?

Well, yes there are! Indeed we can prove the astonishing fact that *every* sufficiently well-stated mathematical problem can be reduced to a question about Life! Those apparently trivial problems about Life histories can be arbitrarily difficult!

Here, for instance, is a tricky little problem that kept mathematicians busy from the time Pierre de Fermat proposed it over 300 years ago until Andrew Wiles solved it in the 1990s. Is it possible for a perfect nth power to be the sum of two smaller ones for any n larger than 2? Despite many learned investigations by many learned mathematicians we still don't know! But if you had an infallible way to foretell the destiny of a given Life configuration, you'd be able to answer this question!

The reason is that we can design for you a finite starting pattern P_0 which will fade away completely if and only if there is a way of breaking an nth power into two smaller ones. If you had a mechanical method which would accept as input an arbitrary finite Life pattern P, and is guaranteed to respond with

> FADE, if the rules of Life will eventually cause P to disappear completely, and
> STAY, if not,

then you could apply it to P_0 and settle Fermat's question.

Even better, we could design a pattern P_1 which will tell you what those perfect powers are. If

$$a^n + b^n = c^n$$

is the first solution of Fermat's problem in a certain dictionary order, then eventually P_1 will lead to a configuration in which there are

> a gliders, travelling North-West,
> b gliders, travelling North-East,
> c gliders, travelling South-West,
> n gliders, travelling South-East,

and nothing else at all! We can do the same sort of thing for other mathematical problems.

Making a Life Computer

Many computers have been programmed to play the game of Life. We shall now return the compliment by showing how to define Life patterns that can imitate computers. Many remarkable consequences will follow from this idea.

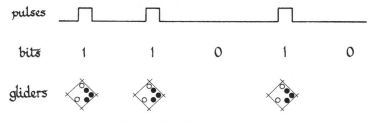

Figure 17. Gliding Pulses.

Good old fashioned computers are made from pieces of wire along which pulses of electricity go. Our basic idea is to mimic these by certain lines in the plane along which gliders travel (Fig. 17). (Because gliders travel diagonally, from now on we'll turn the plane through 45°, so they move across, or up and down, the page.) Somewhere in the machine there is a part called the **clock** which generates pulses at regular intervals and most of the working parts of the machine are made up of logical **gates**, like those drawn in Fig. 18. Obviously we can use Glider Guns as pulse generators. What should we do about the logical gates? Let's study the possible interactions of two gliders which crash at right angles.

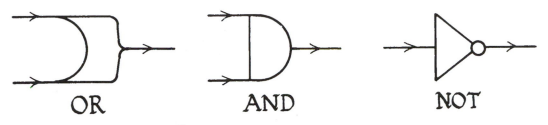

Figure 18. The Three Logical Gates.

When Glider Meets Glider

There are lots of different ways in which two gliders can meet, because there are lots of different possibilities for their exact arrangement and timing. Figure 19 shows them crashing (a) to form a blinker, (b) to form a block, (c) to form a pond, or (d) in one of several ways in which they can annihilate themselves completely. This last may seem rather unconstructive, but these **vanishing reactions** turn out to be surprisingly useful!

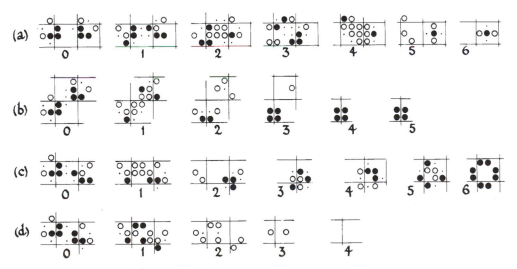

Figure 19. Gliders Crashin' in Diverse Fashion.

How to Make a NOT Gate

We can use a vanishing reaction, together with a Glider Gun, to create a NOT gate (Fig. 20). The input stream enters at the left of the figure, and the Glider Gun is positioned and timed so that every space in the input stream allows just one glider to escape from the gun, while a glider in the stream necessarily crashes with one from the gun in a vanishing reaction (indicated by *). Figure 20 shows the periodic stream

$$1 \quad 1 \quad 0 \quad 1 \quad 1 \quad 0 \quad 1 \quad 1 \quad 0 \quad \ldots$$

being complemented to

$$0 \quad 0 \quad 1 \quad 0 \quad 0 \quad 1 \quad 0 \quad 0 \quad 1 \quad \ldots .$$

being complemented to

1 1 0 1 1 0 1 1 0 ...

0 0 1 0 0 1 0 0 1 ....

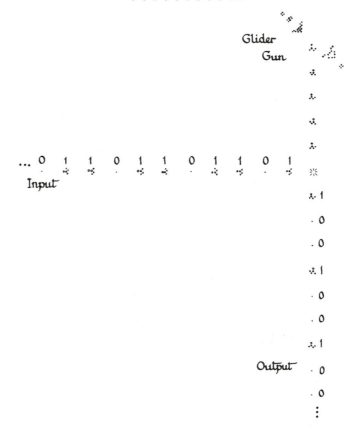

Figure 20. A Glider Gun and a Vanishing Reaction Make a NOT Gate.

Fortunately there are several vanishing reactions with different positions and timings in which the decay is so fast that later gliders from the same gun stream will not be affected (Fig. 21). This means that we can reposition a glider stream arbitrarily by turning it through sufficiently many corners (Fig. 22).

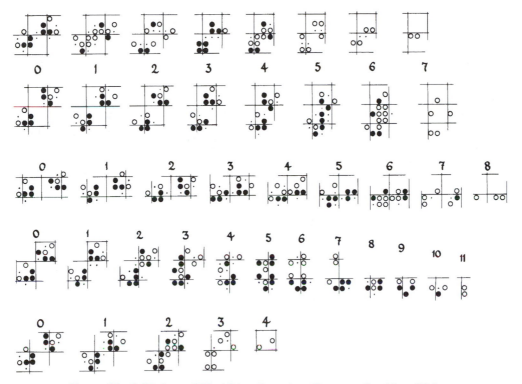

Figure 21. A Variety of Vanishing Reactions Between Crashing Gliders.

The Eater

What else can happen when glider meets glider ? Lots of things! One of them is to make an eater (Fig. 23) and an **eater** can eat lots of things without suffering any indisposition. The eater, which was discovered by Gosper, will be very useful to us; in Fig. 24 you can see it enjoying a varied diet of (a) a blinker, (b) a pre-beehive, (c) a lightweight spaceship, (d) a middleweight spaceship, and (e) a glider . If it attempts a heavyweight spaceship it gets indigestion and leaves a loaf behind; if it tries a blinker in the wrong orientation it leaves a baker's shop!

Sometimes glider streams are embarrassing to have around, so it's especially useful then— it just sits there and eats up the whole stream!

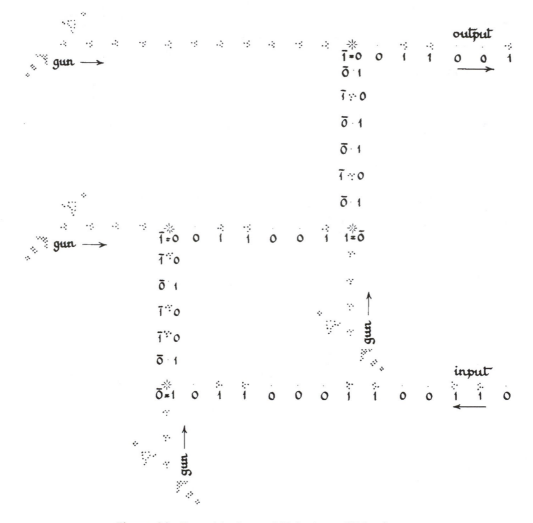

Figure 22. Repositioning and Delaying a Glider Stream.

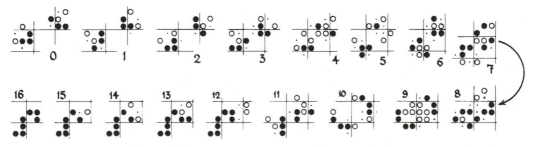

Figure 23. Two Gliders Crash to Form an Eater.

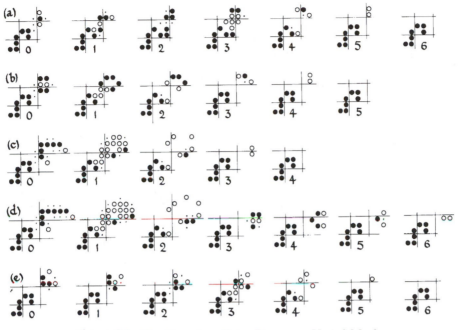

Figure 24. The Voracious Eater Devours a Varied Meal.

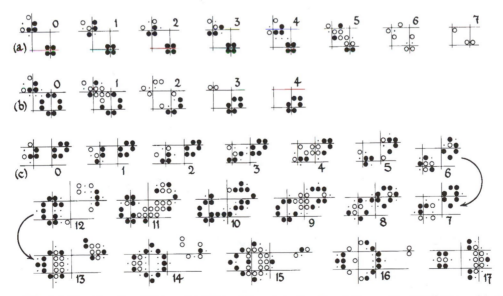

Figure 25. (a) Blockbusting Glider, (b) Glider Dives into Pond and Comes Up With Ship. (c) Glider Crashes into Ship and Makes Part of Glider Gun.

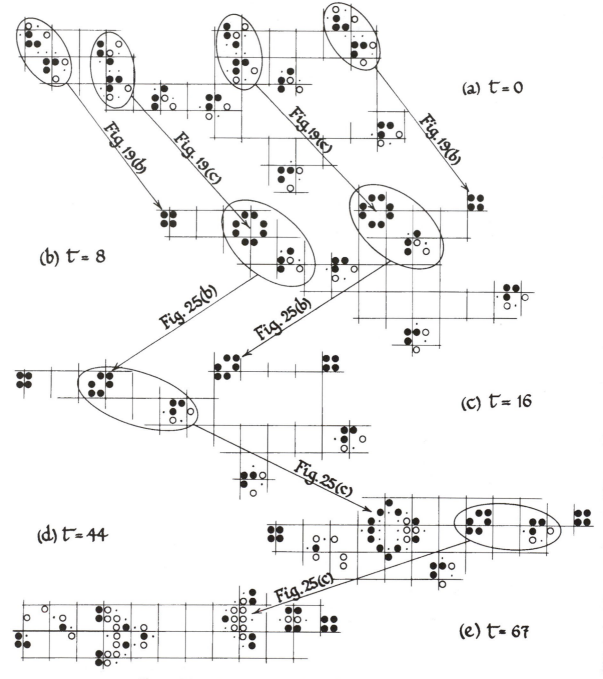

(a) $t = 0$

(b) $t = 8$

(c) $t = 16$

(d) $t = 44$

(e) $t = 67$

Figure 26. Thirteen Gliders Build Their Own Gun.

Gliders Can Build Their Own Guns!

What happens when a glider meets other things? We have seen it get eaten by an eater. It can also annihilate a block (and itself! Fig. 25(a)). But more constructively it can turn a pond into a ship (Fig. 25(b)) and a ship into a part of the glider gun (Fig. 25(c)). And since gliders can crash to make blocks (Fig. 19(b)) and ponds (Fig. 19(c)), they can make a whole glider gun! The 13 gliders in Fig. 26(a) do this in 67 generations. Figures 26(b,c,d) show the positions after 8, 16 and 44 generations. The extra glider then slips in to deal with an incipient beehive, and by 67 generations (Fig. 26(e)) the gun is in full working order and launches its first glider 25 generations later.

The Kickback Reaction

Yet another very useful reaction between gliders is the **kickback** (Fig. 27(a)) in which the decay product is a glider travelling along a line closely parallel to one of the original ones, but in the opposite direction. We think of this glider as having been *kicked back* by the other one. Figure 27(b) shows our notation for the kickback.

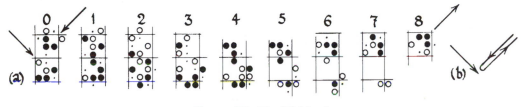

Figure 27. The Kickback.

All the working parts of our computer will be moving glider streams, meeting in vanishing and kickback reactions. The only static parts will be glider guns and eaters (indicated by G and E in the figures).

Thinning a Glider Stream

The glider streams that emerge from normal guns are so dense that they cannot interpenetrate without interfering. If we try to build a computer using streams of this density we couldn't allow any two wires of this kind to cross each other, so we'd better find some way to reduce the pulse rate.

In Fig. 28 the guns G_1 and G_2 produce normal glider streams in parallel but opposite directions. But there is a glider g which will travel West until at A it is kicked East by a glider from the G_1 stream. The timing and phasing are such that at B it will be kicked back towards A again, so that it repeatedly "loops the loop", removing one glider from each of the two streams per cycle. After this every Nth glider is missing from each of these streams. We don't want the G_1 stream, so we feed it into an eater, but we feed the G_2 stream into a vanishing reaction with a stream from a third gun G_3. Every glider from G_2 now dies, but every Nth one from G_3 escapes through a hole in the G_2 stream! So the whole pattern acts as a **thin**

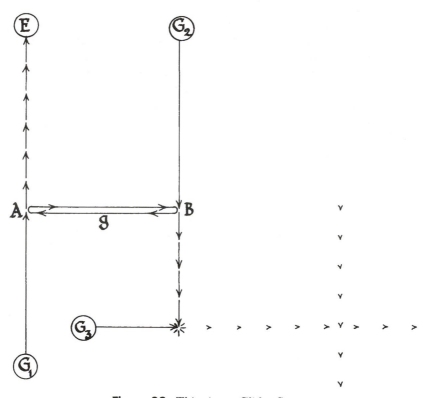

Figure 28. Thinning a Glider Stream.

gun, producing just one Nth as many gliders as the normal gun. To get the phasing right, N must be divisible by 4, but it can be arbitrarily large and so we can make an arbitrarily thin stream. Now two such streams can cross without interacting as in the right hand part of Fig. 28, provided things are properly timed. So from now on we can use the word **gun** to mean an arbitrarily thin gun. Perhaps a thinning factor of 1000 will make all our constructions work.

Building Blocks for Our Computer

In Fig. 29 we see how to build logical gates using only vanishing reactions (we've already seen the NOT gate in more detail in Fig. 20). But there's a problem! The output streams from the AND and OR gates are *parallel* to the input, but the output stream from the NOT gate is at *right angles* to the input. We need a way to turn streams round corners without complementing them, or of complementing them without turning them round corners. Fortunately the solution to our *next* problem automatically solves *this* one.

The new problem is to provide several copies of the information from a given glider stream, and we found it a hard problem to solve. To get some clues, let's see what happens when we use one glider to kick back a glider from a gun stream.

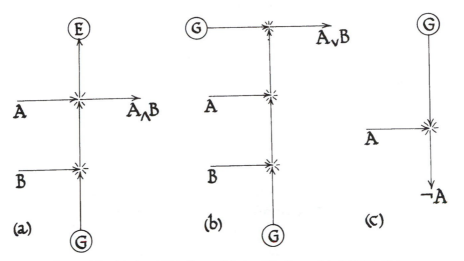

Figure 29. (a) An AND Gate. (b) An OR Gate. (c) A NOT Gate.

We suppose that the gun stream, the **full stream**, produces a glider every 120 generations (a quarter of the original gun density; $N = 4$ in the previous section). Then it turns out that when we kick back the first glider, the effect is to remove just three gliders from the stream! This happens as follows:

(i)   The first glider is kicked back (Fig. 27) along the full stream.
(ii)   The second glider crashes into the first, forming a block (somewhat as in Fig. 19(b)).
(iii) The third glider annihilates the block (Fig. 25(a)).
(iv)   All subsequent gliders from the full stream escape unharmed.

We can use this curious behavior as follows. Suppose that our information-carrying stream operates at one tenth, say, of the density of the full stream, so that the last 9 of every 10 places on it will be empty, while the first place might or might not be full. If we use 0 for a hole and block the places in tens, our stream looks like

$$\ldots 000000000D \; 000000000C \; 000000000B \; 000000000A \rightarrow$$

We first feed it into an OR gate with a stream of type

$$\ldots 00000000g0 \; 00000000g0 \; 00000000g0 \; 00000000g0 \rightarrow$$

the g's denoting gliders that are definitely present. The result is a stream

$$\ldots 00000000gD \; 00000000gC \; 00000000gB \; 00000000gA \rightarrow$$

in which every information-carrying place is definitely followed by a glider g. This stream is used to kick back a full stream whose gliders are numbered:

$$\ldots \; \ldots\ldots\ldots.. \; \ldots\ldots\ldots \; X987654321 \; X987654321 \rightarrow$$

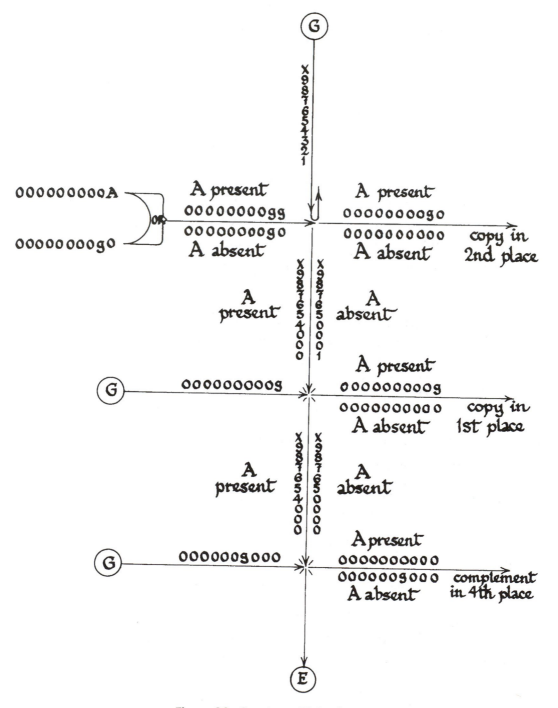

Figure 30. Copying a Glider Stream.

If glider A is *present*, it will obliterate gliders 1, 2 and 3 of the full stream and the following glider g can escape in the confusion. But if A is *absent*, then full stream glider 1 escapes and gliders 2, 3, 4 are removed instead by the following glider g. So the stream which emerges is definitely empty except for the second of every ten places and these places carry a copy of the input stream. The original full stream now manages to carry the information *twice*, in the first and fourth digits of each block, the first digit carrying the *complemented* version (which has not been turned through a right angle). By feeding this stream into vanishing reactions with suitably thin streams we can recover the original stream either complemented or not, and freed from undesirable accompanying gliders! Figure 30 shows these techniques in action.

From here on it's just an engineering problem to construct an arbitrarily large finite (and very slow!) computer. Our engineer has been given the tools—let him finish the job! We know that such computers can be programmed to do many things. The most important ones that we will want it to do involve emitting sequences of gliders at precisely controlled positions and times.

Auxiliary Storage

Of course the engineer will probably have designed an internal memory for our computer using circulating delay lines of glider streams. Unfortunately this won't be enough for the kind of problem we have in mind, and we'll have to find some way of adjoining an *external* memory, capable of holding arbitrarily large numbers. To build this memory, we'll need an additional static piece (the block).

Had Fermat's problem been still unsolved, we might ask the computer to compute

$$a^n + b^n \text{ and } c^n$$

for *all* quadruples (a, b, c, n) in turn and stop when it finds a quadruple for which

$$a^n + b^n = c^n.$$

We don't know how big a, b, c and n might get, and they'll almost certainly get too large even to be written in the internal memory.

So we're going to adjoin some auxiliary storage registers, each of which will store an arbitrarily large number. Figure 31 shows the general plan. Each register contains a block, whose distance from the computer (on a certain scale) indicates the number it contains. In the figure, register A contains 3, B contains 7, C contains 0 and D contains 2. When the contents of a register is 0, the block is just inside the computer. All we have to do is to provide a way for the computer to

increase	the contents of a register by 1,
decrease	the contents of a register by 1, and
test	whether the contents are 0.

Fortunately each of these can be accomplished by a suitable fleet of gliders. One such fleet is off to increase register B by one! And another glider is about to discover that register C contains 0.

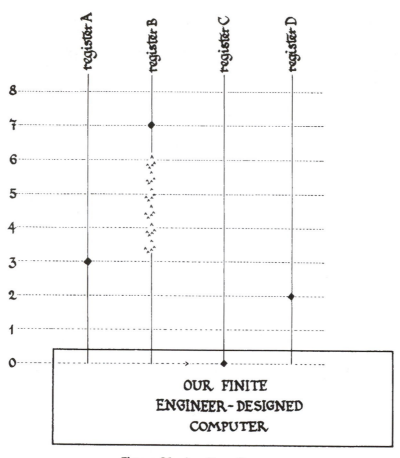

Figure 31. Auxiliary Storage.

How We Move Blocks

To find these fleets we studied the six possible glider-block crashes. One of them does indeed bring the block in a bit, but unfortunately by a knight's move. However the block can be brought back onto the proper diagonal by repeating the process with a reflected glider on a parallel course. The combined effect of this pair of gliders is to pull the block back three diagonal places (Fig. 32).

Unfortunately there is no single glider-block crash which moves the block further away, but there is a crash which produces the arrangement of 4 beehives we call a honey farm , and two of these four are slightly further away, and so we can send in second, third and fourth gliders to annihilate three of the beehives, and then a fifth glider which converts the remaining beehive back into a block. The total effect again pushes the block off the proper diagonal, but a second team of five gliders will restore this, resulting in a block just one diagonal place further out! (Fig. 33).

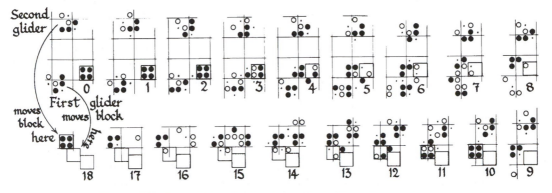

Figure 32. Two Gliders Pull a Block Back Three Diagonal Places.

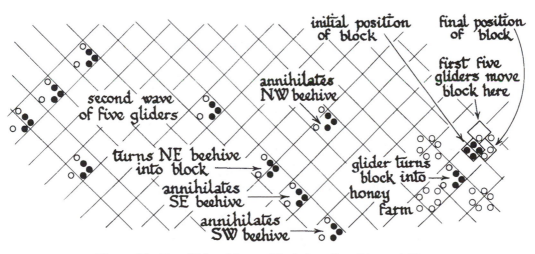

Figure 33. Ten Gliders Move a Block Just One Diagonal Place.

We therefore choose a diagonal distance of 3 to represent a change of 1 in a register and can decrease the contents of a register using a pair of gliders, or increase it using 3 flotillas of 10.

Apart from the difficulty discussed in the next section we have now finished the work, for Minsky has shown that a finite computer, equipped with memory registers like the ones in Fig. 31, can be programmed to attack arbitrarily complicated mathematical problems.

A Little Difficulty

But now comes the problem. Every glider in our finite computer has at some time been produced by a glider gun, so how could we arrange to send those gliders along closely parallel, but distinct paths? Surely one gun would have to fire right through another (Fig. 34)? Our technique of **side-tracking** uses three computer controlled guns G_1, G_2, G_3 as in Fig. 35. These are programmed to emit gliders exactly when we want them to.

Figure 34. How Can One Gun Fire Through Another?

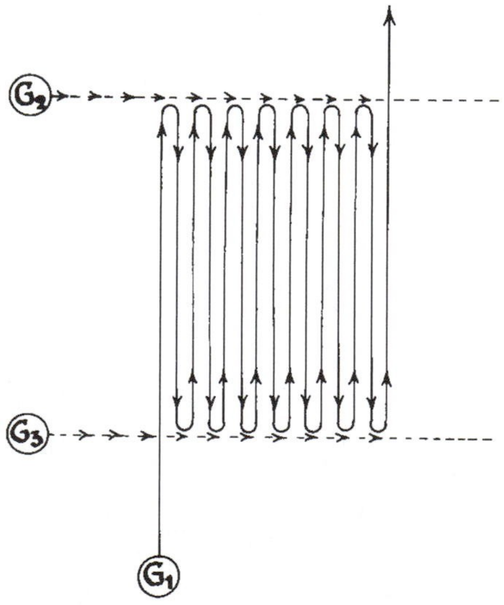

Figure 35. Side-tracking.

Firstly, G_1 emits a glider g travelling upwards,

Secondly, G_2 emits a glider at just the right time to kick g back downwards,

Thirdly, G_3 kicks g back up again,

and so on, alternately, until at a suitable time G_2 fails to fire and g is released. By controlling the number of times G_2 and G_3 fire, the same guns can be used to send a succession of gliders along distinct parallel paths.

Mission Completed—Will Self-Destruct

Side-tracking can be used for a much more spectacular juggling act! We can actually program our computer to throw a glider into the air *and* bring it back down again. In Fig. 36, G_1, G_2, G_3 behave as before and can be programmed to arrange that a glider g ends up travelling Eastwards arbitrarily far above the ground. But G_4 has been arranged to emit a glider which will be kicked back down by g. We could even arrange to kick it back up again, then down again, then up again,... as suggested by the dotted lines in Fig. 36.

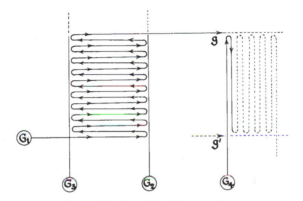

Figure 36. Double Side-tracking.

Using such techniques we can design a program for our computer which will send large numbers of gliders far out into space and then turn them round so that they head back towards the computer along precisely defined tracks (Fig. 37).

Now comes the clever part. Figures 38(a), 38(b) and 25(a) show that the eaters, the guns' moving parts, and blocks, can all be destroyed by aiming suitably positioned gliders from behind their backs. If the computer is cleverly designed we can even destroy it completely by an appropriate configuration of gliders!

Here's the idea. We design the computer so that every glider emitted by a gun or circulating in a loop would, if not deflected by meeting other gliders, be eventually consumed by an appropriately placed eater. Then we design our attacking force of gliders to shoot the computer down, guns first. After each gun is destroyed we wait until any gliders it has already emitted have percolated through the system and either been destroyed by other gliders, or swallowed by eaters, before attacking the next gun. When all the guns are destroyed we shoot down the eaters and blocks.

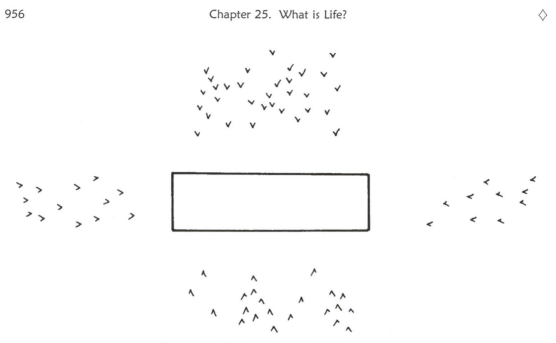

Figure 37. Self-attack from All Directions!

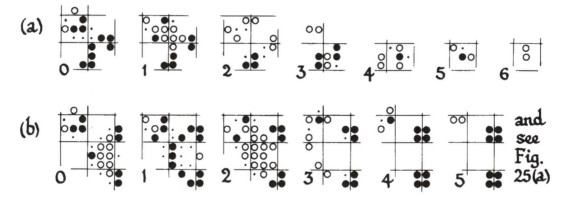

Figure 38. (a) The Eater Eaten! (b) The Gun Gunned Down!

The whole process requires some care. Each gun G_i must have a matching eater E_i, and G_i and E_i lie in a strip of the plane which contains no other static parts of the computer (Fig. 39). The gliders g_1, g_2, g_3,... with which we shoot down a given gun can be arbitrarily widely spaced in time provided they come in along the right tracks. Moreover we can arrange to shoot down the successive guns, eaters and blocks after increasingly long intervals of time.

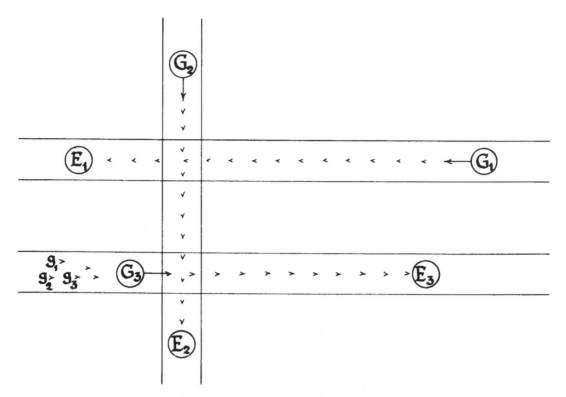

Figure 39. Arranging Destroyable Guns.

However, it *can* be done! We intend to use it like this. Program the computer to look for a solution of an arbitrarily hard problem, such as Fermat's. If it never finds a solution it will just go on forever. However, if it *does* find a solution we instruct it to throw into the air a precisely arranged army of gliders, then reduce all its storage numbers to zero (this brings all the blocks inside the computer), switch off, and await its fate. The attacking glider army, of course, is exactly what's needed to obliterate the computer, leaving no trace. It's important to realize that a *fixed* computer can be programmed to produce many different patterns of gliders and in particular the one required to kill itself. The information about this glider pattern can be held by the numbers in the memory of the computer and not in the computer's design.

Since mathematical logicians have proved that there's no technique which guarantees to tell when arbitrary arithmetical problems have solutions, there's no technique that's guaranteed to tell even when a Life configuration will fade away completely. The kind of computer we have simulated is technically known as a *universal machine* because it can be programmed to perform any desired calculation. We can summarize our result in this answer to our chapter heading:

LIFE IS UNIVERSAL!

Extras

Life is Still Being Lived!

The game of life is very much alive on the Internet, where a search will soon reveal many web pages devoted to it, which have found many new things. We shall only mention Dean Hickerson's page: http://www.math.ucdavis.edu/~dean/life.html and Mark Niemiec's extended counts of still-lifes (i.e., stable objects) and flip-flops (i.e., period 2 oscillators):

size	still-lifes	pseudo still-lifes	flip-flops	pseudo flip-flops
3	0	0	1	0
4	2	0	0	0
5	1	0	0	0
6	5	0	3	0
7	4	0	0	0
8	9	1	1	0
9	10	1	1	0
10	25	7	1	1
11	46	16	1	2
12	121	55	6	14
13	240	110	3	17
14	619	279	20	46
15	1353	620	29	78
16	3286	1645	98	225
17	7773	4067	199	484
18	19044	10843		
19	45759	27250		
20	112243	70637		
21	273188	179011		
22	672172	462086		
23	1646147	1184882		

Life Computers Can Reproduce!

Eaters and guns can be made by crashing suitable fleets of gliders, so it's possible to build a computer simply by crashing some enormously large initial pattern of gliders. Moreover, we can design a computer whose sole aim in Life is to throw just such a pattern of gliders into the air. In this way one computer can give birth to another, which can, if we like, be an exact copy of the first. Alternatively, we could arrange that the first computer eliminates itself after giving birth; then we would regard the second as a reincarnation of the first.

> There are Life patterns which behave
> like self-replicating animals.

> There are Life patterns which move
> steadily in any desired rational
> direction, recovering their initial
> form exactly after some fixed
> number of generations.

Genetic Engineering

We've now shown that among finite Life patterns there is a very small proportion behaving like self-replicating animals. Moreover, it is presumably possible to design such patterns which will survive inside the typical Life environment (a sort of primordial broth made of blocks, blinkers, gliders, ...). It might for instance do this by shooting out masses of gliders to detect nearby objects and then take appropriate action to eliminate them. So one of these "animals" could be more or less adjusted to its environment than another. If both were self-replicating and shared a common territory, presumably more copies of the better adapted one would survive and replicate.

Whither Life?

From here on is a familiar story. Inside any sufficiently large random broth, we expect, *just by chance*, that there will be some of these self-replicating creatures! Any particularly well adapted ones will gradually come to populate their territory. Sometimes one of the creatures will be accidentally modified by some unusual object which it was not programmed to avoid. Most of these modifications, or **mutations**, are likely to be harmful and will adversely affect the animal's chances of survival, but very occasionally, there will be some *beneficial* mutations. In these cases the modified animals will slowly come to predominate in their territory, and so on. There seems to be no limit to this process of evolution.

> It's probable, given a large enough
> Life space, initially in a random state,
> that after a long time, intelligent
> self-reproducing animals will emerge and
> populate some parts of the space.

This is more than mere speculation, since the earlier parts are based on precisely proved theorems. Of course, "sufficiently large" means very large indeed, and we can't prove that "living" animals of any kind are likely to emerge in any Life space we can construct in practice.

It's remarkable how such a simple system of genetic rules can lead to such far-reaching results. It may be argued that the small configurations so far looked at correspond roughly to the molecular level in the real world. If a two-state cellular automaton can produce such varied and esoteric phenomena from these simple laws, how much more so in our own universe?

Analogies with real life processes are impossible to resist. If a primordial broth of amino-acids is large enough, and there is sufficient time, self-replicating moving automata may result from transition rules built into the structure of matter and the laws of nature. There is even the very remote possibility that space-time itself is granular, composed of discrete units, and that the universe, as Edward Fredkin of M.I.T. and others have suggested, is a cellular automaton run by an enormous computer. If so, what we call motion may be only simulated motion. A moving particle in the ultimate microlevel may be essentially the same as one of our gliders, appearing to move on the macrolevel, whereas actually there is only an alteration of states of basic space-time cells in obedience to transition rules that have yet to be discovered.

References and Further Reading

Clark C. Abt , "Serious Games: The Art and Science of Games that Simulate Life", The Viking Press, 1970.

Michael A. Arbib, Simple self-reproducing universal automata. Information and Control, 9 (1966) 177-189.

E.R. Banks , Information Processing and Transmission in Cellular Automata, Ph.D. thesis, M.I.T, 71:01:15.

E.F. Codd , "Cellular Automata", Academic Press, New York and London, 1968.

Martin Gardner , Mathematical Games, Sci. Amer. 223 #4 (Oct. 1970) 120-123; 223 #5 (Nov. 1970) 118; 223 #6 (Dec. 1970) 114; 224 #1 (Jan. 1971) 108; 224 #2 (Feb. 1971) 112-117; 224 #3 (Mar. 1971) 108-109; 224 #4 (Apr. 1971) 116-117; 225 #5 (Nov. 1971) 120-121; 226 #1 (Jan. 1972) 107; 233 #6 (Dec. 1975) 119.

M.J.E. Golay , Hexagonal parallel pattern transformations, IEEE Trans. Computers C18 (1969)733-740.

Chester Lee , Synthesis of a cellular universal machine using the 29-state model of von Neumann, Automata Theory Notes, Univ. of Michigan Engg. Summer Conf., 1964.

Marvin L. Minsky , "Computation: Finite and Infinite Machines", Prentice-Hall, Englewood Cliffs, N.J., 1967.

Edward F. Moore , Mathematics in the biological sciences, Sci. Amer. 211 #3 (Sep. 1964) 148-164.

Edward F. Moore , Machine models of self-reproduction, Proc. Symp. Appl. Math. 14, Amer. Math. Soc. 1962,17-34.

Edward F. Moore , John Myhill , in Arthur W. Burks (ed.) "Essays in Cellular Automata", University of Illinois Press, 1970.

C.E. Shannon , A universal Turing machine with two internal states, in C.E. Shannon and J. McCarthy (eds.) "Automata Studies", Princeton University Press, 1956.

Alvy Ray Smith , Cellular automata theory, Tech. Report No. 2, Digital Systems Lab., Stanford Electronics Labs., Stanford Univ., 1969.

J.W. Thatcher , Universality in the von Neumann cellular model, Tech. Report 03105-30-T, ORA, Univ. of Michigan, 1964.

A.M. Turing , Computing machinery and intelligence, Mind, 59 (1950) 433-460.

Robert T. Wainwright (editor) Lifeline ; a quarterly newsletter for enthusiasts of John Conway's game of Life, 1-11, Mar., Jun., Sep., Dec. 1971, Sep., Oct., Nov., Dec. 1972, Mar, Jun, Sep. 1973.

Glossary

$\mathbf{A} = \mathbf{ace} = \{0 \,|\, \mathbf{tiny}\}$, 357
$\bar{\mathbf{A}} = -\mathbf{ace} = \{\mathbf{miny} \,|\, 0\}$, 359
$\mathbf{A}- = \{\mathbf{on} \,|\, \mathbf{A} \,||\, 0\}$, 359
$\mathbf{A}+ = \{0 \,||\, \bar{\mathbf{A}} \,|\, \mathbf{off}\}$, 359
$\aleph_0$, aleph-zero, 329
Δ, also, slow join, 286
$\wedge$, and, join, 278
$\lceil \ \rceil$, "ceiling," least integer not less than, 485
$(x)_n, \{x|-y\}_n$, Childish Hackenbush values, 238
$a < b >$, class a and variety b, 363
$\bar{1}\clubsuit = \{\clubsuit \,|\, 0\}$, clubs, 359
$\clubsuit = 0\clubsuit = \{1\clubsuit \,|\, 0\}$, clubs, 359
$1\clubsuit = \{\mathbf{deuce} \,|\, 0\}$, clubs, 359
● ● ○ ●, Col and Snort positions, 47, 146
$\gamma°$, degree of loopiness, 361
$2\clubsuit = \{0|\mathbf{ace}\} = \mathbf{ace} + \mathbf{ace} = \mathbf{deuce}$, 357
$\bar{1}\diamondsuit = \{\bar{\mathbf{J}}|0\}$, diamonds, 359
$\diamondsuit = 0\diamondsuit = \{\mathbf{ace} \,|\, \bar{1}\diamondsuit\}$, diamonds, 359
$1\diamondsuit = \{0|\diamondsuit\}$, diamonds, 359
$\Downarrow = \{\downarrow *|0\} = \downarrow + \downarrow$, double-down, 68, 69, 71
$\Uparrow = \{0|\uparrow *\} = \uparrow + \uparrow$, double-up, 68, 69, 71
$\Uparrow * = \{0|\uparrow\} = \uparrow + \uparrow + *$, double-up star, 71
$\downarrow = \{*|0\}$, down, 64
$\downarrow_2 = \{\uparrow *|0\}$, down-second, 235, 236
$\downarrow_3 = \{\uparrow + \uparrow^2 + *|0\}$, down-third, 235, 236
$\downarrow abc \cdots, \downarrow abc \cdots$, 261
$\mathsf{\lor}$, downsum, 336, 357

$\mathbf{dud} = \{\mathbf{dud}|\mathbf{dud}\}$, deathless universal draw, 337
$e = 2.7182818284\ldots$, base of natural logarithms, 610
ϵ, epsilon, small positive number , 328
$\varepsilon = \mathbf{over}$, 674, 675, 688, 689, 691–702
$\doteqdot$, equally uppity, 242, 247, 249
$\lfloor \ \rfloor$, "floor," greatest integer not greater than, 51, 311, 485
$||0$, fuzzy, 29, 32, 37
G, general game, 28–31
$G \,||\, 0, G$ fuzzy 2nd player wins, 29
$G < 0, G$ negative, R wins, 29
$G > 0, G$ positive, L wins, 29
$G = 0, G$ zero, 1st player wins, 29
$G + H$, sum of games, 31
G^L, (set of) L option(s), 31
G^R, (set of) R option(s), 31
$\mathcal{G}(n)$ nim-value, 82
$G. \uparrow = \{G^L. \uparrow + \Uparrow * \ \ | \ \ G^R. \uparrow + \Downarrow *\}$, 247, 255, 258
$>$, greater then, 32
$\geq$ greater than or equal, 32
$\rhd$, greater than or incomparable, 32, 35
$\gtrdot\!\cdot$, at least as uppity, 242, 246
$\frac{1}{2} = \{0|1\}$, half, 7, 20
$\bar{1}\heartsuit = \{\heartsuit|0\}$, hearts, 359
$\heartsuit = 0\heartsuit = \{1\heartsuit|\bar{\mathbf{A}}\}$, hearts, 359
$1\heartsuit = \{0 \,|\, \mathbf{joker}\}$, hearts, 359

hi = {**on** || 0|**off**}, 355
hot = {**on**|**off**}, 355, 710
||, incomparable, 35
$\infty = \mathbb{Z}||\mathbb{Z}|\mathbb{Z}$, 329, 334, 391
$\pm\infty = \infty| - \infty = \mathbb{Z}|\mathbb{Z} = \int^{\mathbb{Z}} *$, 334
$\infty \pm \infty = \infty|0 = \mathbb{Z}|0$, 334
$\infty + \infty = 2.\infty = \mathbb{Z} || \mathbb{Z}|0$, double infinity, 334
$\infty_0. \infty_2$, unrestricted tallies, 314
$\infty_{abc...}$, 387–395
$\infty_{\beta\gamma\delta...}$, 333
$\int$, integral, 167–180
J = {0|**Ā**+} = **ace** $\wedge$(−**ace**) = **joker**, 358
J̄ = {**Ā** − |0} = **ace** $\vee$(−**ace**) = −**joker**, 359
L, Left, 2
LnL, LnR, RnR, positions in Seating games, 44, 45, 132, 133, 260
<, less than, 32
≤ less than or equal, 32
◁, less than or incomparable, 32, 35
lo = {**on**|0 || **off**}, 355
𝔇, loony, 397, 407
$\gamma, \gamma^*, \gamma^+, \gamma^-$, loopy games, 335
$s\&t$ loopy games, 336
$-1 = \{\ |0\}$, minus one, 19
$-_{\mathbf{on}}$ = {**on**|0 || 0}, **miny**, 353
$-_{\frac{1}{4}} = \{\frac{1}{4}|0 || 0\}$, miny-a-quarter, 134
$-_x = \{x|0 || 0\}$, miny-x, 126
$\overset{*}{\times}$ nim-product, 475–477
$\overset{*}{+}$ nim-sum, 58
off = { |**off**}, 336–340, 357, 674, 710
$\omega = \{0, 1, 2, \ldots | \ \}$, 329–333
$\omega + 1 = \{\omega| \ \}$, omega plus one, 329–333
$\omega \times 2 = \{\omega, \omega+1, \omega+2, \ldots | \ \} = \omega+\omega$, 329–333
$\omega^2 = \{\omega, \omega \times 2, \omega \times 3, \ldots | \ \} = \omega \times \omega$, 329–333
on = {**on**| }, 336–341
$1 = \{0| \ \}$, one, 7, 19
1/**on** = **over**, 341, 673, 674, 686
1 **over**, 674, 677, 681, 686
ono = {**on**|0}, 355–356
oof = {0|**off**}, 355–357
over = {0|**over**} = $\frac{1}{\mathbf{on}}$, 341, 674, 688, 696–695
$\pi = 3.141592653\ldots$, pi, 328–329, 610
$\pm1 = \{1| - 1\}$, plus-or-minus-one, 120–122
$(\pm1). \uparrow = \{\Uparrow\uparrow | \Downarrow\}$, 247

$\Downarrow\hspace{-0.3em}\mid = \{\Downarrow * |0\} = 4. \downarrow$, quadruple-down, 347
$\Uparrow\hspace{-0.3em}\mid = 4. \uparrow$, quadruple-up, 71, 347, 693, 706

$\frac{1}{4} = \{0|\frac{1}{2}\}$, quarter, 6, 20
$\hat{\frac{1}{4}} = \frac{1}{4}. \uparrow = \{\Uparrow *|1\frac{1}{2}. \downarrow +*\}$, quarter-up, 236
$\frac{1}{4}* = \frac{1}{4}. \uparrow +* = \{\Uparrow |1\frac{1}{2}. \downarrow\}$, quarter-up-star, 236
R, Right, 2
$\frac{*}{2} = \{*, \uparrow | \downarrow *, 0\}$, semi-star, 370
$\hat{\frac{1}{2}} = \frac{1}{2}. \uparrow = \{\Uparrow *| \downarrow *\}$, semi-up, 236, 247
$\frac{1}{2}* = \frac{1}{2}. \uparrow +* = \{\Uparrow | \downarrow\}$, semi-up-star, 236
$\hat{\frac{3}{2}} = 1\frac{1}{2}. \uparrow = \{*|*\}$, sesqui-up, 236
sign(), 348–350
|, ||, |||, . . . slash, slashes, . . . separate L and R options, 6–7, 128–129, 366
$2\spadesuit = \{\bar{\mathbf{A}}|0\}$, spades, 359
$\bar{1}\spadesuit = \{0|\bar{2}\spadesuit\}$, spades, 359
$\spadesuit = 0\spadesuit = \{0|\bar{1}\spadesuit\}$, spades, 359
$1\spadesuit = 0|\spadesuit$, spades, 359
✩ , far star, remote star, 230–232, 244–251
$* = \{0|0\}$, star, 38
$*2 = \{0, *|0, *\}$, star-two, 41
$*n = \{0, *, \ldots, *(n-1)|0, *, \ldots, *(n-1)\}$, star-$n$, 41
$*\alpha$, star-alpha, 333
$\hat{*} = *. \uparrow = \{\Uparrow *| \downarrow *\}$, starfold-up, 236, 248
$*\bar{n}$, all nimbers except $*n$, 397
$*n \rightarrow .$, all nimbers from $*n$ onwards, 397
$\odot = 0* \rightarrow$, sunny, 397–401, 404
$\uparrow_{abc...} = - \downarrow^{abc...}$, superstars, 261
$2_1 3_0$, tallies, 300–326
$\frac{3}{4} = \{\frac{1}{2}|1\}$, three-quarters, 17
$+_{\mathbf{on}}$ = {0|**oof**} = {0||0|**off**} = **tiny**, 353, 357
$+_{\frac{1}{4}} = \{0 || 0| - \frac{1}{4}\}$, tiny-a-quarter, 126
$+_2 = \{0 || 0| - 2\}$, tiny-two, 126
$+_x = \{0 || 0| - x\}$, tiny-x, 126
tis = {**tisn**| } = 1&0, 342, 374
tisn = { |**tis**} = 0&−1, 342, 374
$(l, r)(l, r)_c$, Toads-and-Frogs positions, 127, 136, 368, 375, 376
$\Downarrow = \{\Downarrow *|0\} = 3. \downarrow$, treble-down, 71, 347
$\Uparrow = \{0| \Uparrow *\} = 3. \uparrow$, treble-up, 71, 347, 693
$3\spadesuit = \{0|\mathbf{deuce}\} = \mathbf{trey} = \mathbf{ace} + \mathbf{deuce}$, 357
△, triangular number, 253
$\overset{*}{\cup}$, ugglies, ugly product, 483–487
$2 = \{1| \ \}$, two, 7, 19
under = {**under**|0} = −**over**, 341
$\vee$,**union**, or, 300

$\uparrow = \{0|*\}$, up, 64, 71, 261
$\uparrow^{\alpha}$, up-alpha, 341
$\uparrow^2 = \{0|\downarrow *\}$, up-second, 235, 236
$\uparrow* = \{0, *|0\}$, up-star, 65, 229
$\uparrow^3 = \{0|\downarrow + \downarrow_2 +*\}$, up-third, 235, 236
$\uparrow_{abc...}$, 261
upon $= \{$**upon**$|*\}$, 341, 375

upon$* = \{0, $**upon**$ * |0\}$, 341
$|\wedge$, upsum, 336, 357
$\triangledown$, ur, urgent number, 312
$[a|b|c|\ldots]_k$, Welter function, 506–514
$\mathbb{Z} = \{\ldots, -2, -1, 0, 1, 2, \ldots\}$, the set of integers, 334
$0 = \{\ \ |\ \ \} = 0* = 0. \uparrow$, zero, 7, 41

Index to Volumes 1–4

money, 171
moneybag, 492
Monopoly, 15
moon, 379, 903–906
   age of, 905
   new, 906
   paschal full, 906
Moore and More, 534
Moore, Eliakim H., 427, 539
Moore, Edward F., 939, 960
Moore, Thomas E., 607
Moore's Nim$_k$, 427, 533
Morelles, 737
Morgenstern, Oskar, 539
moribundity, 598
   equation, 601
Morra, Three-Finger, 15
Morris, Lockwood, 225
Morton, Davis, 640
mosaic, 936
Moser, Leo, 667, 732, 742, 743, 768
motley, 469, 477, 483
Mott-Smith, Geoffrey, 502, 768
mountain, 7
   purple, 197, 198
move 14, 40
   abnormal, 325
   alternating, 46, 47
   bad, 16, 547, 818
   bonus, 405
   chance, 14
   complimenting, 379, 405–407, 541, 552
   consecutive, 405
   darkening, 677, 678, 685
   entailing, 379, 396–405
   equitable and excitable, 161
   exit, 829–830, 838
   five-eights of a, 20
   futile, 654
   good, 16, 196, 397
   half, 4, 19, 20
   horse, 406
   hotter, 173
   illegal, 320–322, 404, 636
   inward, 677

   legal, 404
   lightening, 677, 678, 685
   loony, 322, 398, 400, 411, 561–564, 576–580
   non-entailing, 397–398, 400
   non-suicidal, 322, 562
   normal, 325
   outward, 677
   overriding, 312, 314, 317, 319, 320, 326
   pass, 281, 283, 284, 286, 289, 292, 293, 294, 338, 352, 355
   plausible, 654
   predeciding, 312, 320, 321
   quarter, 6, 20
   repainting, 343
   reversible, 55, 56, 60, 62–64, 66, 70, 71, 75, 77, 126, 212, 213, 415
   reversible misère, 415
   reverting, 425
   strategic, 652
   stupid, 636
   suiciding, 312, 320
   sunny, 397–401, 404, 407, 411
   tactical, 652
   temperature-selected, 132
   three-quarter, 17
   trailing, 402
   worthwhile, 213–216
move set, 515–517
Movie, Magic, 853–854
Mr. Cutt and Mr. Shortt, 745, 746
Mrs. Grundy, 310
Mühle, 737
Müller, Martin, 18, 188, 762, 764
multiples of up, 71, 242, 247, 256, 258
multiples, fractional and non-integral, 236
multiplying pegs, 812–816
multum in parvo, 937
Munro, Ian, 226
Murray, H.J.R., 668, 737, 768
Muscovites, 666
museums, 705
   F&G, 705–708
musical series, 96, 117
mutation, 562, 959
   harmless, 562
   killing, 562
Myhill, John, 960